Berry Antioxidants in Health and Disease

Special Issue Editor
Dorothy Klimis-Zacas

MDPI

Special Issue Editor
Dorothy Klimis-Zacas
Department of Food Science and Human Nutrition,
School of Food and Agriculture
University of Maine
USA

Editorial Office
MDPI AG
St. Alban-Anlage 66
Basel, Switzerland

This edition is a reprint of the Special Issue published online in the open access journal *Antioxidants* (ISSN 2076-3921) in 2016 (available at: http://www.mdpi.com/journal/antioxidants/special_issues/berry_antioxidants).

For citation purposes, cite each article independently as indicated on the article page online and as indicated below:

Author 1; Author 2; Author 3 etc. Article title. *Journalname*. **Year**. Article number/page range.

ISBN 978-3-03842-348-5 (Pbk)
ISBN 978-3-03842-349-2 (PDF)

Table of Contents

About the Guest Editor

Dorothy Klimis-Zacas, PhD. FACN is Professor of Clinical Nutrition and cooperating Graduate Faculty, School of Biomedical Sciences at the University of Maine. She is also cooperating professor of Nutrition and Dietetics at Harokopio University, Athens, Greece and at the Department of Food, Environmental and Nutritional Sciences at the University of Milan, Italy. She was awarded a senior Fulbright Fellowship to the Hellenic School of Public Health, Athens, Greece and two Fulbright Specialist awards to the University of Milan, Italy, Department of Food, Environmental and Nutritional Sciences. Additionally, she was the recipient of the Fondazione Cariplo Fellowship to direct research on the use of biosensors in exploring dietary approaches for chronic disease prevention at the University of Milan.

Dr. Klimis-Zacas has been involved in biomedical research exploring the role of trace minerals and dietary bioactives and functional foods on chronic diseases such as obesity, cardiovascular disease and the metabolic syndrome, including basic and clinical investigations. Her recent investigations examine the role of berries on attenuating co-morbidities associated with the metabolic syndrome such as obesity-induced inflammation, dyslipidemia and insulin resistance. Her applied investigations involve cross-cultural studies that utilize dietary interventions to reduce cardiovascular risk in populations both in the United States and the Mediterranean region.

Dr. Klimis-Zacas is the editor of *Manganese in Health and Disease* and *Nutritional Concerns for Women* CRC Press, and has acted as editor-in-chief of *Annual Editions in Nutrition* and as member of several editorial boards including *the Journal of Nutritional Biochemistry* and board member of the International Society of Trace Mineral Research in Humans (ISTERH). Dr. Klimis-Zacas is a member of many professional societies dedicated to promoting health and preventing disease including The American Society for Nutrition, The American College of Nutrition, The International Atherosclerosis Society, The American Academy of Nutrition and Dietetics, The Italian Society of Nutrition, The Hellenic Dietetic Association and many others.

Preface to "Berry Antioxidants in Health and Disease"

During the last decade, a high volume of work has been published on the health-promoting effects of berries (e.g., blueberries, cranberries, blackberries, *etc.*) that are rich in antioxidant phytochemicals, polyphenols. Consuming a diet rich in polyphenols has been documented to attenuate the risk of chronic diseases, such as cardiovascular disease, certain cancers, diabetes mellitus, and neurodegenerative disorders. Recent evidence also reveals that the biological effects of polyphenols extend beyond their traditional antioxidant role.

This Special Issue includes 10 peer-reviewed papers, including original research papers and reviews. They present the most recent advances in the role of berry antioxidants, not only in maintaining health but also in preventing and/or reversing disease both in cell culture, animal models and in humans. Additionally, the molecular mechanisms and signaling pathways modulated by berry antioxidants are presented. Chapters include the role of berry antioxidants in whole fruit and leaves on the metabolic syndrome, obesity, diabetes and glucose intolerance, cancer, inflammation, oxidative stress and neuroprotection as well as cardiovascular disease. As a guest editor, I would like to acknowledge the authors of all chapters for their valuable contributions and reviewers for their thoughtful and constructive suggestions and time. Special thanks to the publishing team of *the Antioxidants Journal* for their professionalism, attention to detail and timely completion of this volume.

Dorothy Klimis-Zacas
Guest Editor

antioxidants

MDPI

Review

Berry Fruit Consumption and Metabolic Syndrome

Stefano Vendrame [1], Cristian Del Bo' [2], Salvatore Ciappellano [2], Patrizia Riso [2] and Dorothy Klimis-Zacas [1,*]

[1] School of Food and Agriculture, Food Science and Human Nutrition, University of Maine, Orono, ME 04469, USA; stefano.vendrame@fulbrightmail.org

[2] DeFENS—Department of Food, Environmental and Nutritional Sciences, Division of Human Nutrition, Università degli Studi di Milano, 20133 Milan, Italy; cristian.delbo@unimi.it (C.D.B.); salvatore.ciappellano@unimi.it (S.C.); patrizia.riso@unimi.it (P.R.)

* Correspondence: Dorothy_Klimis_Zacas@umit.maine.edu; Tel.: +1-207-581-3124

Academic Editor: Maurizio Battino
Received: 5 July 2016; Accepted: 20 September 2016; Published: 30 September 2016

Abstract: Metabolic Syndrome is a cluster of risk factors which often includes central obesity, dyslipidemia, insulin resistance, glucose intolerance, hypertension, endothelial dysfunction, as well as a pro-inflammatory, pro-oxidant, and pro-thrombotic environment. This leads to a dramatically increased risk of developing type II diabetes mellitus and cardiovascular disease, which is the leading cause of death both in the United States and worldwide. Increasing evidence suggests that berry fruit consumption has a significant potential in the prevention and treatment of most risk factors associated with Metabolic Syndrome and its cardiovascular complications in the human population. This is likely due to the presence of polyphenols with known antioxidant and anti-inflammatory effects, such as anthocyanins and/or phenolic acids. The present review summarizes the findings of recent dietary interventions with berry fruits on human subjects with or at risk of Metabolic Syndrome. It also discusses the potential role of berries as part of a dietary strategy which could greatly reduce the need for pharmacotherapy, associated with potentially deleterious side effects and constituting a considerable financial burden.

Keywords: berries; Metabolic Syndrome; dietary intervention studies; humans

1. Metabolic Syndrome: General Overview

Metabolic Syndrome is characterized by the simultaneous presence of multiple risk factors, which are a direct or indirect consequence of both insulin resistance and overweight/obesity [1]. Although it is not a disease in itself, this combination of health problems dramatically increases the risk of developing type II diabetes mellitus and cardiovascular disease [2].

The US National Cholesterol Education Program–Adult Treatment Panel III (ATPIII) defines Metabolic Syndrome as the combined occurrence of at least three of the following five risk factors: abdominal obesity (waist circumference $\geq$102 cm in males, $\geq$88 cm in females); high blood triglycerides ($\geq$150 mg/dL); low HDL cholesterol ($\leq$40 mg/dL in males, $\leq$50 mg/dL in females); high diastolic and/or systolic blood pressure ($\geq$130/85 mmHg) and high fasting blood glucose ($\geq$100 mg/L) [2].

Together with these diagnostic parameters, which were selected because of their widespread use in the clinical setting, a plethora of other dysfunctional states are often associated with Metabolic Syndrome. Although they are not used for diagnostic purposes, they significantly contribute to the increased cardiovascular risk and the onset of type II diabetes mellitus [2]. In particular, Metabolic Syndrome is associated with a pro-oxidant, pro-inflammatory, and pro-thrombotic state [3,4].

Chronic, low-grade, systemic inflammation is one of the landmark characteristics of Metabolic Syndrome, leading to unnecessary tissue damage, endothelial dysfunction, thrombosis, insulin resistance, high blood pressure, and all the pathologies related to such risk factors, including cardiovascular disease, diabetes, some forms of cancer, arthritis, neurodegenerative diseases, and many others [3,4].

Fat accumulation and obesity are major underlying causes of chronic inflammation, due to the inherent ability of adipocytes to secrete inflammatory mediators: mainly adipocyte-derived cytokines (called adipokines), but also hormones like leptin [5].

Inflammation and fat accumulation both lead to impairment in glucose metabolism and insulin resistance, and in turn, impaired glucose metabolism exacerbates inflammation and tissue damage, thus sustaining a pro-inflammatory state and endothelial dysfunction [6,7].

Indeed, another landmark characteristic of Metabolic Syndrome is the development of endothelial dysfunction, which is strictly related to oxidative stress and inflammation via the nuclear factor kappa-b pathway [1].

Endothelial dysfunction is characterized by an imbalance between vasoconstrictor and vasodilator responses leading to impaired vascular tone, peripheral vascular resistance, and organ perfusion, and is one of the earliest events in the development of atherosclerotic lesions. Metabolic Syndrome is often associated with impaired endothelium-dependent vasodilation, likely due to insufficient NO production or bioavailability, [8] and with exaggerated vasoconstriction, due to an increased production of vasoconstricting mediators [9].

2. The Role of Berries in the Modulation of Metabolic Syndrome

Berries represent a variety of small fruits characterized by the red, purple, and blue color. The most common berries are: blueberry, bilberry, cranberry, blackberry, raspberry, black, white or red currant, and strawberry. Minor berries include: lingonberry, cloudberry, elderberry, honeyberry, whortleberry, and chokeberry.

Berries are consumed both as fresh product as well as processed foods (i.e., juices, beverages, jams, freeze-dried). They contain high levels of polyphenols including flavonoids (anthocyanins, flavonols, and flavanols), condensed tannins (proanthocyanidins), hydrolyzable tannins (ellagitannins and gallotannins), phenolic acids (hydroxybenzoic and hydroxycinnamic acids, chlorogenic acid), stilbenoids and lignans [10,11]. Their concentration varies according to species, genotype, growing and post-harvesting conditions [12].

Anthocyanins (ACNs) are probably the main bioactive compounds that characterize berries with pelargonidin, cyanidin, delphinidin, petunidin, peonidin, and malvidin the most predominant ACN compounds. They are found mainly in the external layer of the pericarp. ACNs include aglycones–anthocyanidins and their glycosides–anthocyanins. They differ with regard to the position and number of hydroxyl groups, degree of methylation, type and number of sugar molecules (mono-, di- or tri-glycosides), type of sugars (the most common sugars include glucose, galactose and arabinose) and type and number of aliphatic or aromatic acids (i.e., *p*-coumaric, caffeic, ferulic acid). Among berries, blackcurrants, black elderberries, blackberries and blueberries are particularly rich in ACNs (400 to 500 mg/100 g) [13]. Phenolic acids are represented by hydroxycinnamic acids (i.e., ferulic, caffeic, p-cumaric acids. and caffeoylquinic esters) and benzoic acid derivatives (i.e., gallic acid, salicylic, p-hydroxybenzoic. and ellagic acids). They occur mainly in bound forms as esters or glycosides. Gallic acid and chlorogenic acids are abundant in blueberry (~200 mg/100 g) and blackberry (~300 mg/100 g). Ellagic acid is the main phenolic acid in strawberries (from 300 mg to 600 mg/100 g) where it is present in free form or esterified to glucose in hydrolysable ellagitannins. Blueberries and cranberries are also important sources of ferulic acid, bilberries and blackcurrants of *p*-coumaric and caffeic acid, while chokeberries are rich sources of caffeic, chlorogenic, and neochlorogenic acids [13].

Flavonols, 3-hydroxyflavones and tannins are also widespread in berries. Flavan-3-ols are a complex subclass of polyphenols without glycoside residues and with different levels of polymerization ranging from monomeric, oligomeric, and polymeric forms. Tannins include both condensed non-hydrolysable tannins known as proanthocyanidins, and esters of gallic and ellagic acids called hydrolysable tannins. Tannins play an essential role in defining the sensory properties of fresh fruit and fruit-derived products. They are responsible for the taste and changes in the color of fruit and fruit juice. Moreover, tannins stabilize ACNs by binding to form co-polymers. The amount of flavan-3-ols and proanthocyanidins in chokeberries, blueberries, and strawberries varies from 150 mg to 700 mg/100 g, while in blackberries and raspberries the amount is about 300 mg/100 g [13].

Blackberries, blueberries, and strawberries are rich sources of ellagitannins (up to 600 mg/100 g), followed by chokeberries, cloudberries and red raspberries (~260 mg/100 g), while small quantities of tannins are found in honeyberries [13].

During recent years, a multitude of clinical research studies have focused on the health properties of berries. In particular, increasing attention has been devoted to the role of berries and their components in the modulation of oxidative stress [14], vascular function [15], inflammation, and lipid metabolism [16]. In addition, research studies have explored the role of berries on chronic diseases such as cardiovascular diseases, diabetes, and obesity with promising, albeit preliminary, results. Several studies have also investigated the effect of berry consumption on their ability to modulate/attenuate risk factors associated with Metabolic Syndrome.

A search of the literature on intervention studies investigating the effect of berry consumption in the modulation of Metabolic Syndrome and/or related risk factors was carried out. Abstracts and full texts from human acute and chronic intervention studies were screened. PUBMED, ScienceDirect, and ScholarGoogle databases were searched to identify articles published later than 1 January 2000. The searches used the following terms and text words alone and in combination: "berry", "Metabolic Syndrome", "overweight", "obesity", "hypertension", "hypercholesterolemia", "hyperlipemia", "Type II diabetes", and "humans". Interventions conducted in healthy subjects (not presenting any of the risk factors characterizing Metabolic Syndrome) were excluded. Reference lists of the obtained articles were also searched for related articles. The search was limited to English-language articles. A total of 45 articles were obtained from the database searches and from their reference lists. Four papers were excluded because studies were performed with a mix of fruits and vegetables and other foods in which berries did not constitute the main food [17–20]. Therefore after exclusions, a total of 41 studies were included in the review (Figure 1). Thirty-four explored a medium-long term intervention, four were post-prandial while three investigated both or performed a chronic intervention followed by an acute study. Here, we summarize the main results of the human studies. The results obtained are also reported in Table 1, describing the type of food or supplement, number of intervention days, number of subjects and their characteristics, dose/day of test food, the use of control/placebo food, and the significant findings.

Table 1. Effect of berry consumption on Metabolic Syndrome and associated risk factors.

Berry	Intervention	Participants	Dose	Main findings	References
Blueberry	8-week, randomized, double-blind, placebo-controlled, parallel intervention	Forty-eight postmenopausal women (Blueberry group: BMI 30.1 ± 5.94 kg/m²; age 59.7 ± 4.58 year; Control group: BMI 32.7 ± 6.79 kg/m²; age 57.3 ± 4.76 year) with pre- and stage 1-hypertension	Blueberry group: 480 mL blueberry drink (22 g freeze-dried blueberry powder corresponding to 1 fresh cup blueberries) Control group: 480 mL placebo drink	↓systolic, diastolic blood pressure and brachial-ankle pulse wave velocity ↑NO plasma levels ↑superoxide dismutase activity after blueberry and control group =weight, waist circumference, CRP	Johnson et al. [21]
	6-week randomized, double-blind, placebo-controlled, parallel intervention	Forty-four subjects with metabolic syndrome (Blueberry group: BMI 35.2 ± 0.8 kg/m²; age 55 ± 2 year; Control group: BMI 36.0 ± 1.1 kg/m²; age 59 ± 2 year)	Blueberry group: Smoothie prepared with 45 g blueberry powder Control group: identical smoothie without blueberry bioactives	↑endothelial function =blood pressure and insulin sensitivity	Stull et al. [22]
	6-week randomized, placebo-controlled, crossover intervention	Eighteen male (BMI 24.8 ± 2.6 kg/m²; age 47.8 ± 9.7 year) with CVD risk factors	Wild blueberry group: 250 mL blueberry drink (25 g WB powder, equivalent to 148 g fresh WB) Control group: 250 mL water with sensory characteristics similar to the WB drink	↓Endogenous and oxidatively-induced DNA damage in PBMCs =lipid profile, weight, markers of inflammation and endothelial function, dietary markers, DNA repair activity	Riso et al. [23]
	6-week randomized, double-blinded, placebo-controlled, parallel intervention	Twenty-seven (BMI between 32 and 45 kg/m²; age > 20 year) obese, insulin-resistant subjects	Blueberry group: smoothie prepared with 45 g of blueberry powder (22.5 g twice a day) (equivalent to 2 cups of fresh blueberries) Control group: identical smoothie without blueberry bioactives	↑insulin sensitivity =markers of inflammation, lipid profile and blood pressure	Stull et al. [24]
	8-week, randomized, single-blinded, controlled parallel intervention	Forty-eight subjects with metabolic syndrome (4 males and 44 females; BMI 37.8 ± 2.3 kg/m²; age 50.0 ± 3.0 year)	Blueberry group: 480 mL blueberry drink (50 g freeze-dried blueberries corresponding to 350 g fresh berries) Control group: 480 mL water	↓systolic and diastolic blood pressure ↓plasma ox-LDL, MDA and HNE levels =lipid profile, weight, waist circumference, inflammation markers	Basu et al. [25]
Bilberry	8-week, randomized, controlled, parallel intervention	Twenty-seven subjects (Bilberry group: BMI 31.4 ± 4.7 kg/m²; age 53 ± 6 year; Control group: 32.9 ± 3.4 year; age 50 ± 7 year) with metabolic syndrome	Bilberry group: 200 g of bilberry purée and 40 g of dried bilberries (eq. 200 g of fresh bilberries) Control group: habitual diet. The use of berries was allowed at maximum of 1 dL/day (corresponding to 80 g/day).	↓serum levels of hs-CRP, IL-6, IL-12 and inflammation score. ↓expression of MMD and CCR2 transcripts associated with monocyte and macrophage function associated genes =body weight, glucose, and lipid profile	Kolehmainen et al. [26]
	5-week, randomized, cross-over intervention	Eighty overweight and obese women (BMI 29.6 ± 2.1 kg/m²; age 44.2 ± 6.2 year; 21 subjects meeting metabolic syndrome criteria)	Bilberry group: 100 g fresh bilberries Sea buckthorn group: Sea buckthorn (SB), SB fractions or SB oils (equivalent to 100 g of berries) Control group: none	↓body weight, waist circumference, VCAM-1, TNF-α, adiponectin ↑insulin, GHbA₁C =fat percent, blood pressure, fasting plasma cholesterol, triacylglycerol, ALAT and IL-6 serum levels	Lehtonen et al. [27]
	4-week, randomized, controlled, parallel intervention	Sixty-two subjects (Bilberry group: BMI range 19.9–31.7 kg/m²; age range 34–68 year; Control group: BMI range 17.8–31.5 kg/m²; age range 30–68 year) with CVD risk factors	Bilberry group: 330 mL bilberry juice/day (diluted to 1 L using tap water) Control group: 1 L of water	↓serum levels of CRP, IL-6, IL-15, TNF-α, MIG =markers of antioxidants status and oxidative stress	Karlsen et al. [28]

Table 1. Cont.

Berry	Intervention	Participants	Dose	Main findings	References
Cranberry	4-week, placebo-controlled double-blind, crossover intervention	Thirty-five abdominally obese men (age 45 ± 10 year, BMI 28.3 ± 2.4 kg/m² and a waist circumference ≥90 cm) with metabolic (n = 13) and without metabolic syndrome (n = 22)	Cranberry group: 500 mL/day of either low-calorie juice (27% juice) Control group: 500 mL/day placebo juice	↓arterial stiffness and global endothelial function =blood pressure, markers of endothelial function	Ruel et al. [29]
	60 days, parallel intervention	Fifty-six subjects (cranberry group: BMI 30.9 kg/m² median, age 51.0 year median; control group: BMI 34.0 kg/m² median, age 48.5 year median) with metabolic syndrome	Cranberry group: 700 mL/day reduced-energy cranberry juice Control group: usual diet	↓serum homocysteine levels, lipoperoxidation, protein oxidation ↑serum folic acid levels =metabolic and inflammatory biomarkers C-reactive protein, TNF-α, IL-1 and IL-6	Simão et al. [30]
	12-week, randomized, double-blind, parallel intervention	Fifty-eight (BMI 28.8 ± 3.6 kg/m², age 54.8 ± 9.1 year) Type II diabetic subjects	Cranberry group: 240 mL/day cranberry juice Control group: 240 mL/day placebo juice	↓glucose, ApoB ↑ApoA-1, PON-1 activity =lipoprotein(a)	Shidfar et al. [31]
	Post-prandial 4-week, randomized, placebo-controlled, cross-over intervention	Fifteen (BMI not reported, age 62 ± 8 year, 13% female) subjects with coronary artery diseases Forty-seven subjects with coronary artery diseases (cranberry group: BMI 30 ± 5 kg/m², age 61 ± 11 year; Control group: BMI 29 ± 4 kg/m², age 63 ± 9 year)	Cranberry group: 480 mL cranberry juice Control group: None Cranberry group: 480 mL/day cranberry juice Control group: 480 mL/day placebo juice	↑flow mediated dilation and lnPAT score as markers of endothelial function =blood pressure, heart rate, brachial diameter, hyperemic flow ↓carotid-femoral pulse wave velocity (a measure of central aortic stiffness), and HDL-cholesterol after cranberry juice =lipid profile, glucose, insulin, HOMA-IR, C-reactive protein, ICAM-1 serum levels, brachial artery flow-mediated dilation, digital pulse amplitude tonometry, blood pressure, and carotid-radial pulse wave velocity	Dohadwala et al. [32]
	8-week, randomized double-blind, placebo-controlled, parallel intervention	Thirty-one (BMI 40.0 ± 7.7 kg/m², age 52.0 ± 8.0 year) female with metabolic syndrome	Cranberry group: 480 mL/day cranberry juice Control group: 480 mL/day placebo drink	↓ox-LDL, MDA & HNE plasma/serum levels ↑Total plasma antioxidant capacity =blood pressure, glucose, plasma lipoprotein-lipid, markers of inflammation	Basu et al. [33]
	Post-prandial cross-over intervention	Thirteen (6 female and 7 male) noninsulin-dependent subjects (age 61.6 ± 2.3 year, BMI 33.25 ± 1.22 kg/m²)	Cranberry group: Group 1: raw cranberry (55 g, 21 cal, 1 g fiber) Group 2: sweetened dried cranberry (40 g, 138 cal; 2.1 g fiber) Group 3: sweetened dried cranberries-less sugars group (40 g; 113 cal; 1.8 g fiber + 10 g polydextrose) Control group: white bread (57 g, 160 cal; 1 g fiber)	↓glycemic and insulinemic response following SDC-LS	Wilson et al. [34]

Table 1. *Cont.*

Berry	Intervention	Participants	Dose	Main findings	References
	12-week intervention (three 4-week intervention with 125, 250 and 500 mL/day cranberry juice)	Thirty (BMI 27.8 ± 3.2 kg/m², age 51 ± 10 year; 9 subjects with metabolic syndrome and 21 without metabolic syndrome) abdominally obese men	Cranberry group: 125, 250 and 500 mL/day cranberry juice; Control group: none	↓ox-LDL following 250 and 500 mL cranberry juice; ↓systolic blood pressure, s-VCAM, ICAM plasma levels following 500 mL cranberry juice; ↓ox-LDL, ICAM plasma levels in subjects with metabolic syndrome following 12-week intervention; ↑HDL cholesterol following 250 and 500 mL cranberry juice; =Total, LDL Apo B cholesterol, triglycerides, diastolic blood pressure, heart beat, E-selectin plasma levels	Ruel et al. [35]
	12-week, randomized, placebo-controlled, double-blind, parallel intervention	Thirty (16 males and 14 females) Type II diabetic subjects (Cranberry group; 9/6 male/female, BMI 26.2 ± 0.7 kg/m², age 65 ± 2 year; Control group: 7/8 male/female, BMI 25.9 ± 1.0 kg/m², age 66 ± 2 year)	Cranberry group: 500 mg/capsule cranberry powder extract, three times a day; Control group: 500 mg/capsule placebo	↓Total cholesterol, Total: HDL cholesterol ratio, LDL cholesterol; =waist circumference, BMI, fasting serum glucose, insulin, HbA1c, HOMA insulin resistance, C-reactive protein, blood pressure, ox-LDL, triglyceride, HDL-cholesterol levels, uric acid	Lee et al. [36]
	Post-prandial intervention	Twelve (6 male and 6 female) type II diabetic subjects (age 65.3 ± 2.3 year, BMI 34.7 ± 1.6 kg/m²)	Cranberry group: *Group 1:* normal calorie cranberry juice (NCCBJ; 27% cranberry juice, *v/v*; 130 Cal/240 mL) *Group 2:* unsweetened low-calorie cranberry juice (LCCBJ; 27%, *v/v* CBJ; 19 Cal/240 mL) Control group: *Group 1:* normal calorie control (NCC; 140 Cal/240 mL) made with dextrose *Group 2:* low-calorie control (LCC; 19 Cal/240 mL)	↓plasma insulin and glycemic response following LCCBJ	Wilson et al. [37]
	12-week intervention (three 4-week intervention with 125, 250 and 500 mL/day cranberry juice)	Thirty (BMI 27.8 ± 3.3 kg/m², age 51 ± 10 year) abdominally obese men	Cranberry group: 125, 250 and 500 mL/day cranberry juice; Control group: none	↓body weight, BMI, waist circumference, waist-to-hip ratio, total:HDL cholesterol apo B, after intervention with 250 and 500 mL cranberry juice; ↑plasma nitrite/nitrate following intervention with 500 mL; ↑plasma antioxidant capacity following 250 and 500 mL cranberry juice; ↑HDL cholesterol following 250 mL cranberry juice; =Total, LDL and VLDL cholesterol	Ruel et al. [38]

Table 1. *Cont.*

Berry	Intervention	Participants	Dose	Main findings	References
	2-week intervention	Twenty-one (BMI 26.9 ± 3.8 kg/m², age 38 ± 8 year) abdominally obese-dyslipidemic men	Cranberry group: 7 mL/kg BW (range 460–760 mL/day cranberry juice) Control group: none	↓BMI, plasma ox-LDL levels ↑Total plasma antioxidant capacity =waist and hip circumference, waist/hip ratio, blood pressure, plasma lipoprotein-lipid, inflammation markers	Ruel et al. [39]
	12-week randomized, controlled, parallel intervention	Twenty-seven Type II diabetic subjects (cranberry group: 14 subjects, 6 women and 8 men, BMI not reported, age 57.9 ± 10.6 year; placebo group: 13 subjects 6 women and 7 men, BMI not reported, age 52.6 ± 13.7 year)	Cranberry group: 6 capsules (equivalent to 240 mL cranberry juice) containing cranberry juice concentrate powder Control group: 6 capsules containing placebo powder	↑insulin levels after placebo treatment =fasting serum glucose, HbA1c, fructosamine, triglyceride, HDL, LDL-cholesterol levels	Chambers & Camire [40]
Raspberry	14-day (4-day black-raspberry intake, post-prandial, randomized, cross-over intervention, wash-out and 4-day black-raspberry intake)	Ten older overweight and obese (BMI, 31.4 ± 2.7 kg/m², age 64.7 ± 6.9 year) males	Berry group: 45 g/day of lyophilized black raspberry powder for 4 days + high fat meal post prandial Control group: No black raspberry + high fat meal post prandial	↓serum IL-6 levels =TNF-α, CRP	Sardo et al. [41]
	12-week, randomized, controlled, parallel intervention	Seventy-seven subjects (berry group: BMI, 26.3 ± 4.3 kg/m², age 58.0 ± 9.2 year; control group: BMI, 25.1 ± 4.0 kg/m², age 60.1 ± 9.5 year) with metabolic syndrome	Berry group: 750 mg/day of black raspberry powder as capsules Control group: 750 mg/day of cellulose, isomalto, and corn powder as capsules	↓total cholesterol level total cholesterol/HDL ratio IL-6, TNF-α ↑flow mediated dilation, adiponectin =serum lipid profile, CRP, ICAM and VCAM	Jeong et al. [42]
Chokeberry	4-week intervention	Twenty-three subjects (BMI, not reported, age 47.5 ± 10.4 year) with hypertension	Berry group: 200 mL/day of polyphenol-rich organic chokeberry juice Control group: none	↓systolic and diastolic blood pressure, heart rate high-frequency power, heart rate very low frequency, standard deviation of normal RR intervals Holter ECG =lipid profile, glucose, CRP, urea, creatinine, Ac. uricum, AST, ALT, markers related to short term heart rate and Holter ECG	Kardum et al. [43]
	4-week intervention	Twenty women (BMI, 36.1 ± 4.4 kg/m², age 53.0 ± 5.4 year) with abdominal obesity	Berry group: 100 mL/day glucomannan-enriched (2 g), chokeberry juice-based Control group: none	↓BMI, waist circumference, systolic blood pressure, serum HDL cholesterol, erythrocytes monounsaturated fatty acids, n6/n3 ratio ↑erythrocytes n3 polyunsaturated fatty acids =erythrocytes saturated and n6 polyunsaturated fatty acids, unsaturation index, diastolic blood pressure, lipid profile, glucose and enzymatic activity (SOD, CAT, GPx)	Kardum et al. [44]
	8-week intervention	Fifty-two subjects (42–65 years old; berry group: 38 subjects, BMI 31.1 ± 3.3 kg/m², control group: 14 healthy subjects; BMI 24.4 ± 1.5 kg/m²) with metabolic syndrome	Berry group: 300 mg/day chokeberry extract Control group: No intervention	↓serum total and LDL cholesterol, triglycerides, coagulation and platelet aggregation parameters =BMI, waist circumference, serum HDL cholesterol	Sikora et al. [45]

Table 1. *Cont.*

Berry	Intervention	Participants	Dose	Main findings	References
	18-week (6-week intervention + 6-week wash-out + 6 week intervention) intervention	Fifty-eight men (BMI 27.7 ± 2.9 kg/m², age 54.1 ± 5.6 year) with mild hypercholesterolemia	Berry group: 150 mL/day chokeberry juice; Control group: none	↓serum total and LDL cholesterol, triglycerides, glucose, homocysteine and fibrinogen, blood pressure; ↑serum HDL₂ cholesterol; =serum HDL and HDL3 cholesterol, hs-CRP, lipid peroxides, uric acid	Skoczyńska et al. [46]
Strawberry	Post-prandial, randomized, controlled, 4-arm, crossover intervention	Twenty-one (BMI 40.2 ± 7.2 kg/m², age 39.8 ± 13.8 year) subjects with abdominal obesity and insulin-resistance	Berry group: Group 1: 10 g freeze-dried whole strawberry powder FDS; Group 2: 20 g FDS; Group 3: 40 g FDS; Control group: 0 g FDS	↓plasma insulin, insulin: incremental increase, glucose: incremental increase, insulin: glucose ratio after 40 g FDS consumption; =plasma glucose, TG, ox-LDL, IL-6	Park et al. [47]
	12-week, randomized, controlled, parallel intervention	Sixty volunteers with CVD risk factors LD-FDS: 15 subjects (14 females/1 male; BMI, 34.5 ± 4.4 kg/m²; age, 50 ± 10 year) HD-FDS: 15 subjects (13 females/2 males; BMI, 38.0 ± 7.1 kg/m²; age, 49 ± 11 year) LD-C: 15 subjects (14 females/1 male; BMI, 37.0 ± 4.4 kg/m²; age, 48 ± 10 year) HD-C: 15 subjects (14 females/2 male; BMI, 35.0 ± 5.2 kg/m²; age 48 ± 10 year)	Berry group: Group 1: low dose freeze-dried strawberries (LD-FDS): 25 g of freeze-dried powder reconstituted in 2 cups (474 mL/day) of water daily (corresponding to 250 g of fresh strawberries); Group 2: high dose freeze-dried strawberries (HD-FDS): 50 g of freeze-dried powder reconstitute in 2 cups (474 mL/day) of water (corresponding to 500 g of fresh strawberries) Control group: Group 1: Low-dose calorie- and fiber-matched control (LD-C) Group 2: high-dose calorie- and fiber-matched control (HD-C)	↓serum total and LDL-cholesterol, derived small LDL particles, MDA and HNE levels following HD-FDS intervention. ↓MDA and HNE following LD-FDS intervention =serum glucose, HbA1c, insulin, HDL, VLDL-cholesterol, HOMA-IR, TG, VACM-1, ICAM-1, hs-CRP	Basu et al. [48]
	6-week, randomized, double-blind, controlled, parallel intervention	Thirty-six diabetic subjects (berry group: n = 19; BMI, 27.36 ± 4.23 kg/m²; age 51.9 ± 8.2 year; control group: n = 17; BMI, 28.58 ± 4.7 kg/m²; age 51.1 ± 13.8 year)	Berry group: 2 cups freeze-dried strawberry (50 g/day); Control group: 2 cups iso-caloric drink with strawberry flavoring	↓diastolic blood pressure =serum TG, total cholesterol, ratio total cholesterol/HDL-cholesterol, systolic blood pressure, anthropometric indices	Amani et al. [49]
	6-week randomized double-blind controlled intervention	Thirty-six subjects (23 females/13 males) with type 2 diabetes (berry group: 19 diabetic subjects; BMI, 27.32 ± 3.26 kg/m², age, 51.88 ± 8.26 year; control group: 17 diabetic subjects; BMI, 28.70 ± 4.24 kg/m², age, 51.17 ± 13.88 year)	Berry group: 50 g/day freeze-dried strawberry (equivalent to 500 g fresh strawberries) Control group: Placebo powder	↓MDA, HbA1c and hs-CRP serum levels ↑total serum antioxidant status =serum glucose levels	Moazen et al. [50]
	7-week double-blind, randomized, cross-over intervention	Twenty obese subjects (BMI between 30 and 40 kg/m², age between 20 and 50 year)	Berry group: Strawberry powder (amount no reported) equivalent to 320 g/day of frozen strawberries. Two servings of strawberry powder mixed as a milkshake, in yogurt, cream cheese, or water-based, sweetened beverage Control group: Milkshake, yogurt, cream cheese, or water-based, sweetened beverage	↓Serum total cholesterol, small HDL particles, LDL size, fibrinogen =serum lipid profile, lipid particle concentrations and size, inflammatory markers and oxidative stress	Zunino et al. [51]

Table 1. *Cont.*

Berry	Intervention	Participants	Dose	Main findings	References
	Post-prandial, randomized, single-blind, placebo-controlled, cross-over intervention	Twenty-six overweight subjects (BMI, 29.2 ± 2.3 kg/m², age, 50.9 ± 15.0 year)	Berry group: High-carbohydrate, moderate-fat meal + strawberry drink (10 g strawberry powder) Control group: High-carbohydrate, moderate-fat meal + placebo drink	↓serum hs-CRP, IL-6 and insulin levels =serum levels of PAI-1, IL-1β, TNF-α, glucose	Ederisinghe et al. [52]
	6-week, randomized, single-blind, placebo-controlled, parallel intervention + post-prandial high carbohydrates high fat meal	Twenty-four overweight and obese subjects (BMI, 29.2 ± 2.3 kg/m²; age, 50.9 ± 15.0 year)	Berry group: 10 g/day strawberry powder in 305 mL water Control group: 10 g/day placebo powder in 305 mL water	=serum levels of glucose, insulin, hs-CRP, IL-6, PAI-1, IL-1β, TNF-α after 6-week intervention ↓serum PAI-1, IL-1β levels following 6-week + post-prandial intervention =serum levels of glucose, insulin, hs-CRP, IL-6, TNF-α after 6-week + post-prandial intervention	Ellis et al. [53]
	6-week, randomized, single-blind, placebo-controlled, crossover intervention After 6 weeks subjects consumed a high fat meal (post-prandial)	Twenty-four hyperlipidemic subjects (14 female, 10 male; BMI, 29.2 ± 2.3 kg/m², age 50.9 ± 15 year)	Berry group: Drinks containing 10 g/serving of freeze-dried strawberry (equivalent to 110 g/day of fresh strawberries) Control group: Placebo drink	↓serum total, LDL and HDL cholesterol, TG after 6 weeks =ox-LDL plasma levels after 6 weeks ↓LDL and HDL cholesterol, TG and ox-LDL plasma levels after post-prandial compared to control group =serum total cholesterol after post-prandial	Burton-Freeman et al. [54]
	8-week intervention	Twenty-seven subjects (BMI: 37.5 ± 2.15 kg/m²; age: 47.0 ± 3.0 year) with metabolic syndrome	Berry group: Four cups of strawberry drink per day (each cup containing 25 g of freeze-dried strawberry powder) Control group: Four cups of water	↓total and LDL-cholesterol, small LDL particles, VCAM-1 =glucose, triglycerides, HDL-cholesterol, lipoprotein particle size and concentrations, blood pressure, and waist circumference, ICAM-1.	Basu et al. [55]
	4-week intervention	Sixteen female (mean BMI, 38.6 ± 2.3 kg/m²; range age, 39–71 year) with metabolic syndrome	Berry group: Two cups of strawberry drink per day (each cup containing 25 g of freeze-dried strawberry powder) Control group: None	↓serum total and LDL-cholesterol, MDA and HNE levels =serum HDL, TG, HDL, VLDL-cholesterol, hs-CRP, adiponectin, glucose, ox-LDL, plasma levels, blood pressure, body weight, waist circumference	Basu et al. [56]
	4-week, randomized, controlled, cross-over intervention	Twenty-eight hyperlipidemic subjects (range BMI, 19.8–32.3 kg/m²; range age, 38–75 years)	Berry group: 454 g/day strawberries Control group: 65 g/day oat bran bread	↓MDA plasma absolute concentration and molar ration of LDL-cholesterol at 4 weeks compared to placebo ↑protein thiols concentration for both the treatments compared to each baseline =plasma levels of conjugated dienes, lipid profile, C-reactive protein, blood pressure, body weight	Jenkins et al. [57]

Table 1. *Cont.*

Berry	Intervention	Participants	Dose	Main findings	References
Whortleberry	4-week, randomized, double-blind, placebo-controlled, parallel intervention	Fifty hyperlipidemic subjects (berry group: twenty-five, 15 females/10 males; BMI, 25.40 ± 1.75 kg/m², age, 48.08 ± 16.39 year; control group: twenty-five hyperlipidemic subjects, 15 females/10 males; BMI, 25.21 ± 2.01 kg/m²; age, 46.36 ± 16.59 year)	Berry group: 2 capsules/day whortleberry capsules / Control group: 2 capsules/day placebo capsules	↓serum total, LDL-cholesterol, TG and plasma MDA levels =serum HDL-cholesterol levels, hs-CRP and BMI	Soltani et al. [58]
Berry mix	12-week, randomized, double-blinded, placebo-controlled intervention	One hundred and thirty-three hypertensive subjects (BMI, 26 ± 3 kg/m², age 62 ± 6 year)	1 Berry group: 500 mL/day (MANA Blue juice: red grapes, cherries, chokeberries and bilberries) / 2 Berry group: 500 mL/day (Optijuice: MANA + polyphenol-rich extract from blackcurrant press-residue) / 3 Control group: 500 mL/day placebo	↓blood pressure	Tjelle et al. [59]
	12-week, randomized, controlled, parallel intervention	Twenty subjects (berry group: BMI, 31.8 ± 4.4 kg/m², age 53.0 ± 6.5 year; control group: BMI, 32.9 ± 3.4 kg/m², age 49.8 ± 7.1 year) with symptoms of metabolic syndrome	Berry group: 300 g/day fresh berries comprising of 100 g of strawberry purée, 100 g of frozen raspberries, and 100 g of frozen cloudberries / Control group: Usual diet	↓serum leptin levels ↑microbiota in both the groups =blood pressure, serum lipid profile, 8-isoprostanes, resistin, TRAP assay	Puupponen-Pimiä et al. [60]
	20-week, randomized, controlled, parallel intervention	Sixty-one female subjects 35–52 years (berry group: BMI 29.3 ± 2.2 kg/m², control group: BMI 29.5 ± 1.8 kg/m²) with metabolic syndrome	Berry group: 163 g/day mix of northern berries (lingonberry, sea buckthorn berry, bilberry and black currant) / Control group: Usual diet	↓plasma ALT levels ↑plasma adiponectin levels =blood pressure, lipid profile, glucose, inflammatory markers, antioxidant capacity	Lehtonen et al. [61]

Legend: BMI: body mass index; NO: nitric oxide; hs-CRP: high sensitive C-reactive protein; ICAM: Intercellular Adhesion Molecule 1; VCAM: Vascular cell adhesion molecule 1; ALT/ALAT: alanine aminotransferase; TRAP: total radical trapping antioxidant parameter; TG: triglycerides; LDL: low density lipoprotein; ox-LDL: oxidized low density lipoprotein; HDL: high density lipoprotein; MDA: malondialdehyde; PBMC: peripheral blood mononuclear cell; HNE: 4-hydroxynonenal; IL-1β: interleukin-1β; IL-6: interleukin-6; TNF-α: tumor necrosis factor-alpha; PAI-1: plasminogen activator inhibitor-1; HOMA-IR: Insulin resistance index; GST: glutathione S-transferase; SOD: superoxide dismutase; GPx: glutathione peroxidase; GHbA₁C: Glycated hemoglobin A₁C; PAT: peripheral arterial function; PONI: paraxonase1; MIG: monokine induced by interferon; MMD: monocyte to macrophage differentiation associated; CCR2: chemokine (C-C motif) receptor 2.

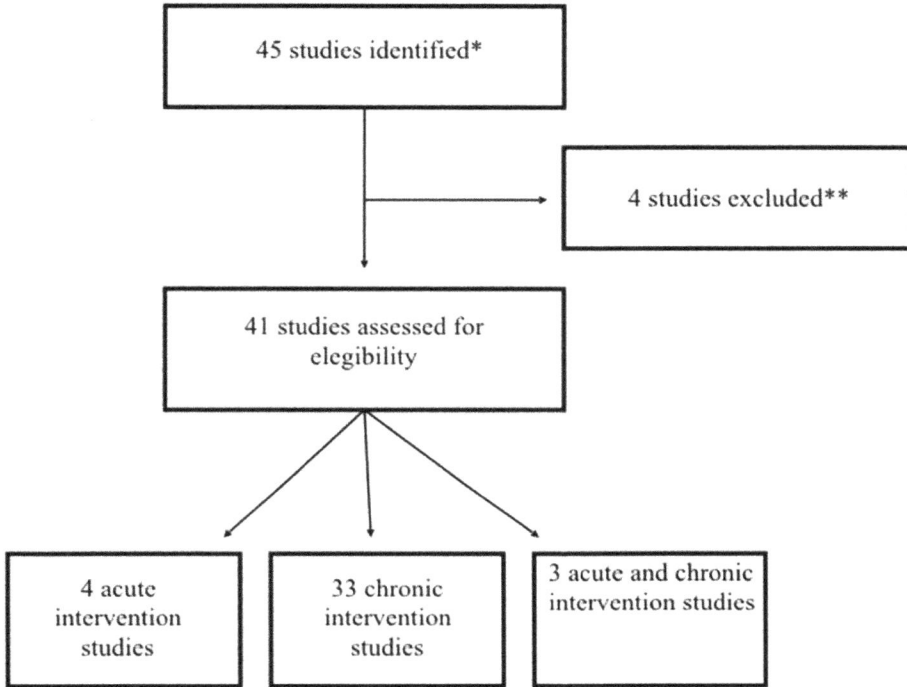

Figure 1. A flow chart highlighting study selection. * Studies were identified according to the following keywords: "berry", "Metabolic Syndrome", "overweight", "obesity", "hypertension", "hypercholesterolemia", "hyperlipemia", "Type II diabetes", and "humans". ** Studies were excluded because they used a mix of fruit and vegetables and other foods in which berries were not the main food.

3. Blueberry

Depending on the size of the shrubs, blueberries are classified as either lowbush ('wild') or highbush: *Vaccinium corymbosum*, a highbush blueberry, is the most commonly cultivated species around the world, while *Vaccinum angustifolium*, native to North America, is the most widespread species of wild blueberry [62]. Anthocyanins are the distinctive and most abundant class of phenolic compounds in blueberries, and wild blueberries contain higher amounts of anthocyanins compared to highbush blueberries. Blueberries also contain significant amounts of phenolic acids, flavonols and flavan-3-ols, as well as manganese [63].

In an 8-week, controlled, randomized intervention, 48 middle-aged, obese subjects with Metabolic Syndrome were given a daily blueberry drink made with 50 g freeze-dried blueberries, resulting in significantly lower systolic and diastolic blood pressure, plasma oxidized LDLs, and malondialdehyde (MDA) and 4-hydroxynonenal (HNE) concentrations. No significant changes were observed for cholesterol and triglyceride levels, inflammatory markers CRP, adiponectin and IL-6, or adhesion molecules ICAM-1 and VCAM-1 [25].

In a 6-week, placebo-controlled, randomized intervention, 27 obese, insulin-resistant subjects received a daily smoothie made with 45 g of blueberry powder, which resulted in improved insulin sensitivity but did not affect markers of inflammation, blood pressure, and blood lipid profile [24]. In a subsequent 6-week intervention with the same experimental design on 44 obese subjects with Metabolic Syndrome, endothelial function measured as peripheral arterial tone, significantly improved with blueberry consumption, while blood pressure and insulin sensitivity did not significantly change [22].

In a 6-week, placebo-controlled, cross-over, randomized intervention on 18 subjects with risk factors for cardiovascular disease receiving a daily blueberry drink made of 25 g wild blueberry powder, endogenous and ex vivo oxidatively-induced DNA damage in blood mononuclear cells was found to be significantly lower, but no changes were observed in blood lipid profile, markers of inflammation, and endothelial function [23].

In an 8-week, placebo-controlled, cross-over, randomized intervention on 48 overweight or obese hypertensive women, administration of a daily drink made with 22 g of blueberry powder resulted in significantly lower systolic and diastolic blood pressure and brachial-ankle pulse wave velocity, and higher plasma levels of nitric oxide, but no changes in the inflammatory marker CRP [21].

In conclusion, the effect of blueberries in the modulation of Metabolic Syndrome-related risk factors has been investigated in five interventions, documenting an effect on blood pressure, oxidative stress, endothelial function, and insulin sensitivity, but not on inflammatory marker levels or blood lipid profile.

4. Bilberry

Bilberries (*Vaccinium myrtillus*), the wild European blueberries, have a phenolic profile and content which is comparable to that of blueberries, and are also exceptionally rich sources of anthocyanins.

In a 4-week, controlled, randomized intervention, 62 subjects with risk factors for cardiovascular disease received 330 mL/day of bilberry juice, resulting in decreased serum concentrations of inflammatory markers CRP, IL-6, IL-15 and TNF-α, while markers of antioxidant status and oxidative stress remained unaffected [28].

When 100 g/day of fresh bilberries were given to 80 overweight or obese women with or without Metabolic Syndrome for 5 weeks, body weight, waist circumference and concentrations of VCAM-1 and TNF-α, but also adiponectin, significantly decreased, insulin and glycated hemoglobin concentrations increased, while fat percent, blood pressure, blood lipids, and IL-6 levels were not affected [27].

In an 8-week, controlled, randomized intervention, in 27 overweight or obese subjects with Metabolic Syndrome receiving 200 g/day of fresh bilberry purée and 40 g/day of dried bilberries, serum concentrations of inflammatory markers hs-CRP, IL-6, IL-12, and LPS decreased, with no changes in blood glucose and lipid profile [26].

Thus, the effect of bilberries on Metabolic Syndrome has not been fully investigated. To our knowledge, only three interventions focused on this berry, finding a consistent effect on reducing inflammation levels.

5. Cranberry

Native to North America, cranberry (*Vaccinium macrocarpon*) is a rich source of different phenolic antioxidant compounds, primarily phenolic acids and flavonoids including anthocyanins, flavonols, and flavan-3-ols [33]. Cranberries also contain resveratrols and the anti-inflammatory acetylsalicylic acid [29].

A randomized, placebo-controlled intervention administering six daily capsules of cranberry powder (equivalent to 240 mL cranberry juice) for 12 weeks to a group of 27 subjects with type 2 diabetes, did not detect any significant effect on blood glucose levels, glycated hemoglobin HbA1c, triglyceride or cholesterol levels [40].

A 2-week intervention on 21 dyslipidemic men with abdominal obesity, reported improved total plasma antioxidant capacity, and decreased BMI and plasma oxidized LDL levels following daily consumption of 7 mL/kg BW of cranberry juice. Blood pressure, plasma lipid profile, and inflammatory markers were not significantly affected [39].

When the effects of three different doses (125, 250, and 500 mL/day) of cranberry juice administered sequentially for four weeks each, were tested on a group of 30 middle-aged men with abdominal obesity, a significant reduction in body weight, BMI, waist circumference, total/HDL-cholesterol ratio and apolipoprotein B, and a significant improvement in plasma total antioxidant capacity were reported after 250 mL and/or 500 mL consumption. A significant decrease in plasma nitrite and nitrate, and increase in HDL cholesterol were reported only after the highest dose. No significant effect was observed at the lowest dose, and no effect on total, LDL and VLDL cholesterol was observed at any dose [38].

The same experimental design was tested on 30 middle-aged abdominally obese men, nine of which with Metabolic Syndrome. HDL cholesterol and oxidized LDLs improved after 250 and 500 mL cranberry juice consumption, while systolic blood pressure and plasma ICAM-1 and VCAM-1 concentrations decreased only at the highest dose, with a stronger effect in the subset of subjects with Metabolic Syndrome. No significant effect was observed at the lowest dose, and no effect on cholesterol, triglycerides, diastolic blood pressure, and plasma E-selectin was observed at any dose [35].

When the post-prandial effect of a single-dose of sweetened cranberry juice was compared to unsweetened cranberry juice in a group of 12 obese subjects with type 2 diabetes, only unsweetened cranberry juice resulted in reduced plasma insulin levels and glycemic response, compared to both a sweetened cranberry juice and an unsweetened control drink [37].

With a similar experimental design, the post-prandial effects of either raw cranberries, sweetened dried cranberries or dried cranberries sweetened with less sugar were tested in 13 obese subjects with type 2 diabetes. The effects of raw cranberries were the lowest on glycemic and insulinemic responses, with the less sweetened, dried cranberries significantly lower compared to the fully sweetened ones [34].

In a 12-week, placebo-controlled, randomized intervention, 30 older overweight subjects with type 2 diabetes received a daily capsule of 500 mg cranberry powder extract, resulting in lower total and LDL cholesterol, and total/HDL-cholesterol ratio. Blood glucose, insulin, and glycated hemoglobin HbA1c levels were unaffected, as well as CRP, blood pressure, oxidized LDLs, triglycerides, and HDL-cholesterol [36].

In a placebo-controlled, randomized intervention, 31 obese women with Metabolic Syndrome were given 480 mL/day of cranberry juice for eight weeks, resulting in reduced circulating levels of oxidized LDLs, malondialdehyde (MDA), and 4-hydroxynonenal (HNE), and improved total plasma antioxidant capacity, while blood pressure, blood lipid and glucose levels, and inflammatory markers did not change [33].

When a single-dose of 480 mL cranberry juice was administered to 15 subjects with coronary artery disease, post-prandial endothelial function measured as brachial artery flow-mediated dilation, it significantly improved (Dohadwala et al., 2011). When the same serving of cranberry juice was given daily to a group of 44 subjects with coronary artery disease in a 4-week, placebo-controlled, cross-over intervention, central aortic stiffness, measured as carotid-femoral pulse wave velocity, significantly decreased, as well as HDL-cholesterol levels. However, no changes were reported for total and LDL cholesterol, triglycerides, CRP, ICAM-1, blood pressure, carotid-radial pulse wave velocity or brachial artery flow-mediated dilation [32].

In a 12-week, placebo-controlled, randomized intervention on 58 middle-aged, overweight or obese subjects with type 2 diabetes, receiving 240 mL/day of cranberry juice, blood glucose, ApoB and ApoA1 concentrations significantly decreased, PON-1 activity increased, while lipoprotein(a) concentrations were unaffected [31].

When 56 overweight or obese subjects with Metabolic Syndrome were given 700 mL/day of low-calorie cranberry juice for four weeks, serum folic acid levels significantly increased while serum homocysteine levels, lipid and protein oxidation significantly decreased. Inflammatory markers CRP, TNF-α, IL-1, and IL-6 did not change [30].

Thirty five abdominally obese men with or without Metabolic Syndrome were given 500 mL/day of low-calorie 27% cranberry juice for four weeks in a placebo-controlled, cross-over intervention. Measures of global endothelial function and arterial stiffness significantly decreased following cranberry consumption compared to baseline, but not compared to post-placebo (Ruel et al., 2013 [29]).

Blood pressure and serum markers of endothelial function ICAM-a, VCAM-1 and E-selectin, as well as oxidized LDLs, did not change [29].

In summary, the effects of cranberries on markers of Metabolic Syndrome have been investigated in eleven interventions, resulting in a consistent effect on reducing oxidative stress, and some beneficial effects on BMI, blood cholesterol levels, and vascular function.

6. Raspberry

Black raspberry (Rubus occidentalis) is a flavonoid-rich member of the Rosaceae family common in Japan, China, and South Korea, where it is traditionally used to treat prostate and urinary diseases. It also contains significant amounts of tyrosol, resveratrol, tannins, and other phenolic acids [42].

In a 12-week, randomized, controlled intervention, a group of 77 Metabolic Syndrome patients received 750 mg/day of black raspberry powder in the form of capsules or a fiber- and sugar-matched control. Raspberry resulted in decreased total serum cholesterol and total/HDL-cholesterol ratio, with no other changes in serum lipid profile. Inflammatory markers IL-6 and TNF-α significantly decreased, and anti-inflammatory adiponectin significantly increased, while CRP, ICAM-1 and VCAM-1 were unaffected. Brachial artery flow-mediated dilation significantly improved [42].

In a controlled, cross-over intervention on a group of 10 older overweight or obese men, the short-term effect of relatively large amounts of lyophilized black raspberry powder (45 g/day for four days) was evaluated in countering the post-prandial inflammatory effects of a high fat meal. Serum IL-6 concentrations significantly decreased, but not TNF-α and CRP concentrations [41].

In conclusion, the impact of raspberry consumption on subjects with risk factor for Metabolic Syndrome has been evaluated in only two interventions, documenting some positive effects on blood cholesterol levels, inflammation and vascular function. Research is needed to investigate more fully the potential of this berry.

7. Chokeberry

Chokeberry (*Aronia melanocarpa*) is a violet-black, strongly-flavored berry, common in areas of North America and Eastern Europe. It is a rich source of polyphenols, particularly anthocyanins present in form of cyanidin-glucosides, but also caffeic acid, quercetin, procyanidins, and other flavonoids [46].

An 18-week intervention on a group of 58 overweight or obese men with mild hypercholesterolemia who were given 150 mL/day of chokeberry juice for two, 6-week long periods separated by a 6-week wash-out, resulted in decreased circulating levels of serum total and LDL cholesterol and increased serum HDL_2 cholesterol. Furthermore, serum triglycerides, glucose, homocysteine, and fibrinogen levels also significantly decreased, as well as blood pressure, while inflammatory markers hs-CRP protein and lipid peroxides were not affected [46].

Eight-week administration of 300 mg/day of chokeberry extract to a group of 52 middle-aged individuals with Metabolic Syndrome, resulted in decreased levels of serum total and LDL cholesterol, with no changes in total HDL cholesterol. Serum triglycerides were also significantly decreased, as well as parameters of platelet aggregation and coagulation [45].

A 4-week intervention on 20 middle-aged, obese women with 100 mL/day of a glucomannan-enriched chokeberry juice resulted in decreased BMI, waist circumference and systolic blood pressure. However, no changes were observed in diastolic blood pressure, blood glucose levels, serum triglycerides, total and LDL cholesterol, with increased levels of HDL cholesterol. Membrane composition of erythrocytes was affected, with an increase in omega-3 polyunsaturated fatty acids and a decrease in monounsaturated fatty acids [44].

Four week administration of 200 mL/day of chokeberry juice to a group of 23 hypertensive subjects resulted in decreased systolic and diastolic blood pressure, but blood lipid profile, glucose and C-reactive protein levels were unaffected [43].

The effect of chokeberries on Metabolic Syndrome has been investigated in four intervention studies, documenting some positive but not consistent effects on blood cholesterol and triglyceride levels, blood pressure, and erythrocyte membrane fluidity. More research is needed to better investigate the effects of this berry.

8. Strawberry

Besides being an important source of vitamin C, strawberries (*Fragaria* × *ananassa*) contain an abundance of various phenolic compounds, including quercetin, ellagic acid, anthocyanins, catechins, and kaempferol [54].

Following a controlled 7-week, randomized, cross-over intervention, 28 hyperlipidemic subjects receiving a serving of strawberries (454 g/day) showed reduced LDL oxidative damage and a reduced LDL-cholesterol molar ratio compared to control. No significant effect was observed for plasma levels of conjugated dienes, lipid profile, C-reactive protein, blood pressure, and body weight [57].

An 8-week randomized, controlled trial on 27 subjects with Metabolic Syndrome receiving two cups of a strawberry beverage (composed of 25 g of freeze-dried strawberry powder each) daily for eight weeks, documented a reduction of total and LDL-cholesterol, small LDL particle size, and VCAM-1. No significant effect was observed for glucose, triglycerides, HDL-cholesterol, blood pressure, waist circumference, and ICAM-1 [55].

A 4-week intervention on 16 obese women with Metabolic Syndrome receiving two daily cups of a strawberry drink (composed of 25 g of freeze-dried strawberry powder each), reported a significant reduction in serum total and LDL-cholesterol levels, with no changes in HDL cholesterol, triglycerides, blood glucose levels and blood pressure. Lipid peroxidation products malondialdehyde (MDA) and 4-hydroxynonenal (HNE) significantly decreased, but not plasma oxidized LDL, and inflammatory markers hs-CRP and adiponectin [56].

A placebo-controlled, randomized, crossover 6-week intervention on 24 middle-aged, overweight or obese hyperlipidemic subjects receiving a daily strawberry drink (composed of 10 g of freeze-dried strawberry powder) did not find any differences in fasting triglycerides, cholesterol or plasma oxidized LDL concentrations. However, when the postprandial response to a high-fat meal challenge was evaluated at the end of the intervention, subjects who were on the strawberry group had significantly lower triglycerides as well as total, LDL and HDL-cholesterol levels [54].

A placebo-controlled, randomized, 6-week intervention on 24 overweight subjects receiving the same strawberry drink, found no changes in serum glucose, insulin, hs-CRP, IL-6, PAI-1, IL-1β, and TNF-α concentrations. However, when a high carbohydrate and fat meal was given to the subjects at the end of the intervention, serum PAI-1, IL-1β levels were significantly lower in the strawberry group, but not serum glucose, insulin, hs-CRP, IL-6, and TNF-α levels [53].

When the post-prandial effect of a single-dose of the same strawberry drink was evaluated on a group of 26 overweight individuals consuming a high-carbohydrate moderate-fat meal, without any pre-intervention, serum hs-CRP, IL-6 and insulin levels were found to be significantly lower, but not PAI-1, IL-1β, TNF-α, and blood glucose concentrations [52].

Following a controlled 7-week, randomized, cross-over intervention, 20 obese individuals receiving two servings of strawberry powder (equivalent to 320 g/day of frozen strawberries) showed a significant reduction in serum total and small HDL-cholesterol concentrations and in LDL particle size. No changes in serum triglycerides and inflammatory markers and oxidative stress were observed, but serum fibrinogen concentrations decreased [51].

When a group of 36 overweight subjects with type 2 diabetes received 50 g/day of freeze-dried strawberry for six weeks in a placebo-controlled, randomized intervention, markers of total serum antioxidant status significantly increased while serum malondialdehyde (MDA), glycated hemoglobin HbA1c, and hs-CRP concentrations significantly decreased, and blood glucose levels remained unaffected [50].

In another controlled, randomized intervention, 36 overweight or obese subjects with type 2 diabetes received the same amount of freeze-dried strawberry for six weeks, resulting in a significantly lower diastolic blood pressure, but with no changes in systolic blood pressure as well as serum triglycerides and cholesterol levels [49].

In a 12-week, randomized, controlled intervention, 60 obese volunteers with risk factors for cardiovascular disease were assigned to either a low-dose or high-dose strawberry treatment (25 or 50 g freeze-dried strawberry powder in water, respectively), or a calorie- and fiber-matched control. At the end of the intervention, malondialdehyde (MDA) and 4-hydroxynonenal (HNE) significantly decreased with both doses, but a significant reduction in serum total and LDL-cholesterol, and derived small LDL particles, was only observed in the high-dose group. Serum glucose, glycated hemoglobin HbA1c, insulin, HDL, VLDL-cholesterol, HOMA-IR score, triglycerides, VCAM-1, ICAM-1, and hs-CRP did not change significantly [48].

When a single serving of freeze-dried strawberry powder at different doses (0, 10, 20 or 40 g) was given to a group of 21 abdominally obese subjects with insulin resistance, a significant reduction in post-prandial plasma insulin concentrations, insulin:glucose ratio, and rate of glucose and insulin increase was only observed after consumption of the highest dose. Interestingly, a significant reduction in plasma oxidized LDLs was observed after the 20 g dose, but not after the 40 g dose. Plasma glucose, triglycerides and IL-6 levels did not change at any dose [47].

Thus, ten studies focused on the role of strawberries on Metabolic Syndrome and its risk factors. Overall, strawberry consumption led to some improvements in oxidative stress, inflammatory status, blood lipid profile, and blood pressure, although these effects were not consistent.

9. Whortleberry

Caucasian whortleberry, or qaraqat (*Vaccinium arctostaphylos*), is a wild berry, common in Western Asia with high anthocyanin content [58]. In a 4-week, placebo-controlled, randomized intervention on 50 hyperlipidemic subjects, two daily capsules of whortleberry extract containing 45 mg of anthocyanins each, resulted in lower serum concentrations of total and LDL cholesterol, triglycerides, and MDA, without changes in HDL-cholesterol and inflammatory marker hs-CRP [47]. Due to the paucity of data regarding this berry, more research is needed to investigate its effects.

10. Berry Mix

Considering the positive results obtained from single-berry studies, some research groups tested the effect of a mix of berries in randomized interventions.

In a 12-week, placebo-controlled study on 133 older, borderline or hypertensive subjects, consumption of 500 mL/day of a juice made from red grapes, cherries, chokeberries, and bilberries resulted in significantly lower systolic blood pressure after six, but not after 12 week consumption , and no effect on diastolic pressure. When only the hypertensive sub-group was considered, the blood pressure lowering effect was more pronounced [59].

In a 12 week intervention on 20 overweight or obese subjects with Metabolic Syndrome risk factors, daily combined consumption of 100 g strawberry purée, 100 g of frozen raspberries, and 100 g of frozen cloudberries, resulted in lower serum leptin levels but did not affect blood pressure, serum lipid profile, resistin, and markers of oxidative stress [60].

In a 20 week intervention on 61 overweight or obese women with Metabolic Syndrome, a daily mix of 163 g of lingonberry, sea buckthorn berry, bilberry, and black currant resulted in higher anti-inflammatory adiponectin plasma concentrations, but did not affect blood pressure, blood lipid or glucose levels, markers of oxidative stress, and other markers of inflammation [61].

More studies investigating the effects of a mix of multiple berries would be useful, since berries are often commercialized in mixes, both as fresh products but above all in juices, smoothies, purees, and other products.

11. Remarks and Conclusions

In the US, more than one in every five adults meets the diagnostic criteria for Metabolic Syndrome [64]. In turn, this condition dramatically increases the risk of developing type II diabetes and suffering acute cardiovascular events, such as heart attacks or strokes [1]. Thus, finding economic and effective ways to prevent and reverse Metabolic Syndrome is of key importance for public health.

Several lines of evidence suggest that diet, together with regular physical activity and avoidance of smoking, is one of the most manageable ways of preventing the development of Metabolic Syndrome in the human population, and at the same time it is also a tool to mitigate the symptoms and decrease the risk of complications in patients who already suffer from this condition.

Furthermore, diet has the potential to greatly reduce the need for pharmacological treatment, which is inevitably associated with harmful side effects and constitutes a considerable financial burden. Being a multifactorial condition, the pharmacological approach to Metabolic Syndrome requires the use of multiple medications, to control a wide array of metabolic abnormalities ranging from dyslipidemia to hypertension, and from hypercholesterolemia to impaired blood glucose control. Thus, the risk of side effects is multiplied by the need of administering multiple medications, and the opportunity to be able to prevent or even control some of the same metabolic dysfunctions with lifestyle—including diet—becomes especially advantageous.

Mounting evidence suggests that consumption of dietary achievable amounts of berries, whose distinctive nutritional characteristic is the abundance of phenolic compounds, has the potential to affect several metabolic abnormalities related to Metabolic Syndrome.

When results of the dietary interventions reviewed in this article are considered together, it appears that the stronger and most recurrent effects of berry consumption lie in their anti-inflammatory and anti-oxidant effects. They also have a tendency to improve lipid profile, lowering total and LDL-cholesterol as well as triglycerides, but this outcome is not observed consistently. It is reasonable to speculate that inter-individual variability as well as interactions with the rest of the diet, are strong enough to mask the effects of berry consumption on lipid profile, and this seems to be confirmed by the observation that effects on blood lipids tend to reach statistical significance in subjects with particularly abnormal baseline values.

Several studies also report positive effects in attenuating blood pressure, especially systolic blood pressure in hypertensive subjects, and some studies document positive effects on markers of endothelial function.

As far as the effects on glucose and insulin metabolism are concerned, results are mixed and tend to be absent, in a few studies even negative. It is possible that in subjects whose glucose and insulin metabolism is already severely impaired, the benefits of phenolic compounds are outweighed by the sugar content of berries, especially when given in form of juices and not with meals.

Another recurring observation in studies testing multiple doses is that there exists a dose-response effect, and some positive outcomes are only observed at higher levels of consumption or with longer interventions.

Considering the overall positive effects of berry consumption that have been reported in several dietary interventions on multiple metabolic abnormalities related to Metabolic Syndrome, it appears that regular berry consumption is a promising strategy to prevent Metabolic Syndrome and its complications, certainly not by itself as a single-bullet solution, but as part of a varied, balanced, and healthy dietary approach promoting health and preventing disease.

Conflicts of Interest: The authors declare no conflicts of interest.

References

1.   Ma, X.; Zhu, S. Metabolic syndrome in the prevention of cardiovascular diseases and diabetes—Still a matter of debate? *Eur. J. Clin. Nutr.* **2013**, *67*, 518–521. [CrossRef] [PubMed]

2. Alberti, K.G.; Eckel, R.H.; Grundy, S.M.; Zimmet, P.Z.; Cleeman, J.I.; Donato, K.A.; Fruchart, J.C.; James, W.P.; Loria, C.M.; Smith, S.C., Jr. Harmonizing the metabolic syndrome: A joint interim statement of the International Diabetes Federation Task Force on Epidemiology and Prevention; National Heart, Lung, and Blood Institute; American Heart Association; World Heart Federation; International Atherosclerosis Society; and International Association for the Study of Obesity. *Circulation* **2009**, *120*, 1640–1645. [PubMed]

3. Samad, F.; Ruf, W. Inflammation, obesity, and thrombosis. *Blood* **2013**, *122*, 3415–3422. [CrossRef] [PubMed]

4. Makki, K.; Froguel, P.; Wolowczuk, I. Adipose Tissue in Obesity-Related Inflammation and Insulin Resistance: Cells, Cytokines, and Chemokines. *ISRN Inflamm.* **2013**, *2013*, 139239. [CrossRef] [PubMed]

5. Lafontan, M. Adipose tissue and adipocyte dysregulation. *Diabetes Metab.* **2014**, *40*, 16–28. [CrossRef] [PubMed]

6. Marti, C.N.; Gheorghiade, M.; Kalogeropoulos, A.P.; Georgiopoulou, V.V.; Quyyumi, A.A.; Butler, J. Endothelial dysfunction, arterial stiffness, and heart failure. *J. Am. Coll. Cardiol.* **2012**, *60*, 1455–1469. [CrossRef] [PubMed]

7. Frueh, J.; Maimari, N.; Homma, T.; Bovens, S.M.; Pedrigi, R.M.; Towhidi, L.; Krams, R. Systems biology of the functional and dysfunctional endothelium. *Cardiovasc. Res.* **2013**, *99*, 334–341. [CrossRef] [PubMed]

8. Hamdy, O.; Ledbury, S.; Mullooly, C. Lifestyle modification improves endothelial function in obese subjects with the insulin resistance syndrome. *Diabetes Care* **2003**, *26*, 2119–2125. [CrossRef] [PubMed]

9. Cardillo, C.; Campia, U.; Bryant, M.B.; Panza, J.A. Increased activity of endogenous endothelin in patients with type II diabetes mellitus. *Circulation* **2002**, *106*, 1783–1787. [CrossRef] [PubMed]

10. Nile, S.H.; Park, S.W. Edible berries: Bioactive components and their effect on human health. *Nutrition* **2014**, *30*, 134–144. [CrossRef] [PubMed]

11. Szajdek, A.; Borowska, E.J. Bioactive compounds and health-promoting properties of berry fruits: A review. *Plant Foods Hum. Nutr.* **2008**, *63*, 147–156. [CrossRef] [PubMed]

12. Manganaris, G.A.; Goulas, V.; Vicente, A.R.; Terry, L.A. Berry antioxidants: Small fruits providing large benefits. *J. Sci. Food Agric.* **2014**, *94*, 825–833. [CrossRef] [PubMed]

13. Zanotti, I.; Dall'Asta, M.; Mena, P.; Mele, L.; Bruni, R.; Ray, S.; Del Rio, D. Atheroprotective effects of (poly)phenols: A focus on cell cholesterol metabolism. *Food Funct.* **2015**, *6*, 13–31. [CrossRef] [PubMed]

14. Del Bo', C.; Martini, D.; Porrini, M.; Klimis-Zacas, D.; Riso, P. Berries and oxidative stress markers: An overview of human intervention studies. *Food Funct.* **2015**, *6*, 2890–2917. [CrossRef] [PubMed]

15. Blanch, N.; Clifton, P.M.; Keogh, J.B. A systematic review of vascular and endothelial function: Effects of fruit, vegetable and potassium intake. *Nutr. Metab. Cardiovasc. Dis.* **2015**, *25*, 253–266. [CrossRef] [PubMed]

16. Vendrame, S.; Klimis-Zacas, D. Anti-inflammatory effect of anthocyanins via modulation of nuclear factor-κB and mitogen-activated protein kinase signaling cascades. *Nutr. Rev.* **2015**, *73*, 348–358. [CrossRef] [PubMed]

17. Hanhineva, K.; Lankinen, M.A.; Pedret, A.; Schwab, U.; Kolehmainen, M.; Paananen, J.; de Mello, V.; Sola, R.; Lehtonen, M.; Poutanen, K.; et al. Nontargeted metabolite profiling discriminates diet-specific biomarkers for consumption of whole grains, fatty fish, and bilberries in a randomized controlled trial. *J. Nutr.* **2015**, *145*, 7–17. [CrossRef] [PubMed]

18. Lankinen, M.; Schwab, U.; Kolehmainen, M.; Paananen, J.; Poutanen, K.; Mykkänen, H.; Seppänen-Laakso, T.; Gylling, H.; Uusitupa, M.; Orešič, M. Whole grain products, fish and bilberries alter glucose and lipid metabolism in a randomized, controlled trial: The Sysdimet study. *PLoS ONE* **2011**, *6*, e22646. [CrossRef] [PubMed]

19. De Mello, V.D.; Schwab, U.; Kolehmainen, M.; Koenig, W.; Siloaho, M.; Poutanen, K.; Mykkänen, H.; Uusitupa, M. A diet high in fatty fish, bilberries and wholegrain products improves markers of endothelial function and inflammation in individuals with impaired glucose metabolism in a randomised controlled trial: The Sysdimet study. *Diabetologia* **2011**, *54*, 2755–2767. [CrossRef] [PubMed]

20. Ali, A.; Yazaki, Y.; Njike, V.Y.; Ma, Y.; Katz, D.L. Effect of fruit and vegetable concentrates on endothelial function in metabolic syndrome: A randomized controlled trial. *Nutr. J.* **2011**, *10*, 72. [CrossRef] [PubMed]

21. Johnson, S.A.; Figueroa, A.; Navaei, N.; Wong, A.; Kalfon, R.; Ormsbee, L.T.; Feresin, R.G.; Elam, M.L.; Hooshmand, S.; Payton, M.E.; et al. Daily blueberry consumption improves blood pressure and arterial stiffness in postmenopausal women with pre- and stage 1-hypertension: A randomized, double-blind, placebo-controlled clinical trial. *J. Acad. Nutr. Diet.* **2015**, *115*, 369–377. [CrossRef] [PubMed]

22. Stull, A.J.; Cash, K.C.; Champagne, C.M.; Gupta, A.K.; Boston, R.; Beyl, R.A.; Johnson, W.D.; Cefalu, W.T. Blueberries improve endothelial function, but not blood pressure, in adults with metabolic syndrome: A randomized, double-blind, placebo-controlled clinical trial. *Nutrients* **2015**, *7*, 4107–4123. [CrossRef] [PubMed]

23. Riso, P.; Klimis-Zacas, D.; del Bo', C.; Martini, D.; Campolo, J.; Vendrame, S.; Møller, P.; Loft, S.; de Maria, R.; Porrini, M. Effect of a wild blueberry (*Vaccinium angustifolium*) drink intervention on markers of oxidative stress, inflammation and endothelial function in humans with cardiovascular risk factors. *Eur. J. Nutr.* **2013**, *52*, 949–961. [CrossRef] [PubMed]

24. Stull, A.J.; Cash, K.C.; Johnson, W.D.; Champagne, C.M.; Cefalu, W.T. Bioactives in blueberries improve insulin sensitivity in obese, insulin-resistant men and women. *J. Nutr.* **2010**, *140*, 1764–1768. [CrossRef] [PubMed]

25. Basu, A.; Du, M.; Leyva, M.J.; Sanchez, K.; Betts, N.M.; Wu, M.; Aston, C.E.; Lyons, T.J. Blueberries decrease cardiovascular risk factors in obese men and women with metabolic syndrome. *J. Nutr.* **2010**, *140*, 1582–1587. [CrossRef] [PubMed]

26. Kolehmainen, M.; Mykkänen, O.; Kirjavainen, P.V.; Leppänen, T.; Moilanen, E.; Adriaens, M.; Laaksonen, D.E.; Hallikainen, M.; Puupponen-Pimiä, R.; Pulkkinen, L.; et al. Bilberries reduce low-grade inflammation in individuals with features of metabolic syndrome. *Mol. Nutr. Food Res.* **2012**, *56*, 1501–1510. [CrossRef] [PubMed]

27. Lehtonen, H.M.; Suomela, J.P.; Tahvonen, R.; Yang, B.; Venojärvi, M.; Viikari, J.; Kallio, H. Different berries and berry fractions have various but slightly positive effects on the associated variables of metabolic diseases on overweight and obese women. *Eur. J. Clin. Nutr.* **2011**, *65*, 394–401. [CrossRef] [PubMed]

28. Karlsen, A.; Paur, I.; Bøhn, S.K.; Sakhi, A.K.; Borge, G.I.; Serafini, M.; Erlund, I.; Laake, P.; Tonstad, S.; Blomhoff, R. Bilberry juice modulates plasma concentration of NF-kB related inflammatory markers in subjects at increased risk of CVD. *Eur. J. Nutr.* **2010**, *49*, 345–355. [CrossRef] [PubMed]

29. Ruel, G.; Lapointe, A.; Pomerleau, S.; Couture, P.; Lemieux, S.; Lamarche, B.; Couillard, C. Evidence that cranberry juice may improve augmentation index in overweight men. *Nutr. Res.* **2013**, *33*, 41–49. [CrossRef] [PubMed]

30. Simão, T.N.; Lozovoy, M.A.; Simão, A.N.; Oliveira, S.R.; Venturini, D.; Morimoto, H.K.; Miglioranza, L.H.; Dichi, I. Reduced-energy cranberry juice increases folic acid and adiponectin and reduces homocysteine and oxidative stress in patients with the metabolic syndrome. *Br. J. Nutr.* **2013**, *110*, 1885–1894. [CrossRef] [PubMed]

31. Shidfar, F.; Heydari, I.; Hajimiresmaiel, S.J.; Hosseini, S.; Shidfar, S.; Amiri, F. The effects of cranberry juice on serum glucose, apoB, apoA-I, Lp(a), and Paraoxonase-1 activity in type 2 diabetic male patients. *J. Res. Med. Sci.* **2012**, *17*, 355–360. [PubMed]

32. Dohadwala, M.M.; Holbrook, M.; Hamburg, N.M.; Shenouda, S.M.; Chung, W.B.; Titas, M.; Kluge, M.A.; Wang, N.; Palmisano, J.; Milbury, P.E.; et al. Effects of cranberry juice consumption on vascular function in patients with coronary artery disease. *Am. J. Clin. Nutr.* **2011**, *93*, 934–940. [CrossRef] [PubMed]

33. Basu, A.; Betts, N.M.; Ortiz, J.; Simmons, B.; Wu, M.; Lyons, T.J. Low-energy cranberry juice decreases lipid oxidation and increases plasma antioxidant capacity in women with metabolic syndrome. *Nutr. Res.* **2011**, *31*, 190–196. [CrossRef] [PubMed]

34. Wilson, T.; Luebke, J.L.; Morcomb, E.F.; Carrell, E.J.; Leveranz, M.C.; Kobs, L.; Schmidt, T.P.; Limburg, P.J.; Vorsa, N.; Singh, A.P. Glycemic responses to sweetened dried and raw cranberries in humans with type 2 diabetes. *J. Food Sci.* **2010**, *75*, H218–H223. [CrossRef] [PubMed]

35. Ruel, G.; Pomerleau, S.; Couture, P.; Lemieux, S.; Lamarche, B.; Couillard, C. Low-calorie cranberry juice supplementation reduces plasma oxidized LDL and cell adhesion molecule concentrations in men. *Br. J. Nutr.* **2008**, *99*, 352–359. [CrossRef] [PubMed]

36. Lee, I.T.; Chan, Y.C.; Lin, C.W.; Lee, W.J.; Sheu, W.H. Effect of cranberry extracts on lipid profiles in subjects with Type 2 diabetes. *Diabet. Med.* **2008**, *25*, 1473–1477. [CrossRef] [PubMed]

37. Wilson, T.; Meyers, S.L.; Singh, A.P.; Limburg, P.J.; Vorsa, N. Favorable glycemic response of type 2 diabetics to low-calorie cranberry juice. *J. Food Sci.* **2008**, *73*, H241–H245. [CrossRef] [PubMed]

38. Ruel, G.; Pomerleau, S.; Couture, P.; Lemieux, S.; Lamarche, B.; Couillard, C. Favourable impact of low-calorie cranberry juice consumption on plasma HDL-cholesterol concentrations in men. *Br. J. Nutr.* **2006**, *96*, 357–364. [CrossRef] [PubMed]

39. Ruel, G.; Pomerleau, S.; Couture, P.; Lamarche, B.; Couillard, C. Changes in plasma antioxidant capacity and oxidized low-density lipoprotein levels in men after short-term cranberry juice consumption. *Metabolism* **2005**, *54*, 856–861. [CrossRef] [PubMed]
40. Chambers, B.K.; Camire, M.E. Can cranberry supplementation benefit adults with Type 2 diabetes? *Diabetes Care* **2003**, *26*, 2695–2696. [CrossRef] [PubMed]
41. Sardo, C.L.; Kitzmiller, J.P.; Apseloff, G.; Harris, R.B.; Roe, D.J.; Stoner, G.D.; Jacobs, E.T. An open-label randomized crossover trial of lyophilized black raspberries on postprandial inflammation in older overweight males: A pilot study. *Am. J. Ther.* **2016**, *23*, e86–e91. [CrossRef] [PubMed]
42. Jeong, H.S.; Hong, S.J.; Lee, T.B.; Kwon, J.W.; Jeong, J.T.; Joo, H.J.; Park, J.H.; Ahn, C.M.; Yu, C.W.; Lim, D.S. Effects of black raspberry on lipid profiles and vascular endothelial function in patients with metabolic syndrome. *Phytother. Res.* **2014**, *28*, 1492–1498. [CrossRef] [PubMed]
43. Kardum, N.; Milovanović, B.; Šavikin, K.; Zdunić, G.; Mutavdžin, S.; Gligorijević, T.; Spasić, S. Beneficial effects of polyphenol-rich chokeberry juice consumption on blood pressure level and lipid status in hypertensive subjects. *J. Med. Food* **2015**, *18*, 1231–1238. [CrossRef] [PubMed]
44. Kardum, N.; Petrović-Oggiano, G.; Takic, M.; Glibetić, N.; Zec, M.; Debeljak-Martacic, J.; Konić-Ristić, A. Effects of glucomannan-enriched, aronia juice-based supplement on cellular antioxidant enzymes and membrane lipid status in subjects with abdominal obesity. *Sci. World J.* **2014**, *2014*, 869250. [CrossRef] [PubMed]
45. Sikora, J.; Broncel, M.; Markowicz, M.; Chałubiński, M.; Wojdan, K.; Mikiciuk-Olasik, E. Short-term supplementation with *Aronia melanocarpa* extract improves platelet aggregation, clotting, and fibrinolysis in patients with metabolic syndrome. *Eur. J. Nutr.* **2012**, *51*, 549–556. [CrossRef] [PubMed]
46. Skoczyñska, A.; Jêdrychowska, I.; Porêba, R.; Affelska-Jercha, A.; Turczyn, B.; Wojakowska, A.; Andrzejak, R. Influence of chokeberry juice on arterial blood pressure and lipid parameters in men with mild hypercholesterolemia. *Pharmacol. Rep.* **2007**, *59*, 177–182.
47. Park, E.; Edirisinghe, I.; Wei, H.; Vijayakumar, L.P.; Banaszewski, K.; Cappozzo, J.C.; Burton-Freeman, B. A dose-response evaluation of freeze-dried strawberries independent of fiber content on metabolic indices in abdominally obese individuals with insulin resistance in a randomized, single-blinded, diet-controlled crossover trial. *Mol. Nutr. Food Res.* **2016**, *60*, 1099–1109. [CrossRef] [PubMed]
48. Basu, A.; Betts, N.M.; Nguyen, A.; Newman, E.D.; Fu, D.; Lyons, T.J. Freeze-dried strawberries lower serum cholesterol and lipid peroxidation in adults with abdominal adiposity and elevated serum lipids. *J. Nutr.* **2014**, *144*, 830–837. [CrossRef] [PubMed]
49. Amani, R.; Moazen, S.; Shahbazian, H.; Ahmadi, K.; Jalali, M.T. Flavonoid-rich beverage effects on lipid profile and blood pressure in diabetic patients. *World J. Diabetes* **2014**, *5*, 962–968. [CrossRef] [PubMed]
50. Moazen, S.; Amani, R.; Homayouni Rad, A.; Shahbazian, H.; Ahmadi, K.; Taha Jalali, M. Effects of freeze-dried strawberry supplementation on metabolic biomarkers of atherosclerosis in subjects with type 2 diabetes: A randomized double-blind controlled trial. *Ann. Nutr. Metab.* **2013**, *63*, 256–264. [PubMed]
51. Zunino, S.J.; Parelman, M.A.; Freytag, T.L.; Stephensen, C.B.; Kelley, D.S.; Mackey, B.E.; Woodhouse, L.R.; Bonnel, E.L. Effects of dietary strawberry powder on blood lipids and inflammatory markers in obese human subjects. *Br. J. Nutr.* **2012**, *108*, 900–909. [CrossRef] [PubMed]
52. Edirisinghe, I.; Banaszewski, K.; Cappozzo, J.; Sandhya, K.; Ellis, C.L.; Tadapaneni, R.; Kappagoda, C.T.; Burton-Freeman, B.M. Strawberry anthocyanin and its association with postprandial inflammation and insulin. *Br. J. Nutr.* **2011**, *106*, 913–922. [CrossRef] [PubMed]
53. Ellis, C.L.; Edirisinghe, I.; Kappagoda, T.; Burton-Freeman, B. Attenuation of meal-induced inflammatory and thrombotic responses in overweight men and women after 6-week daily strawberry (*Fragaria*) intake. A randomized placebo-controlled trial. *J. Atheroscler. Thromb.* **2011**, *18*, 318–327. [CrossRef] [PubMed]
54. Burton-Freeman, B.; Linares, A.; Hyson, D.; Kappagoda, T. Strawberry modulates LDL oxidation and postprandial lipemia in response to high-fat meal in overweight hyperlipidemic men and women. *J. Am. Coll. Nutr.* **2010**, *29*, 46–54. [CrossRef] [PubMed]
55. Basu, A.; Fu, D.X.; Wilkinson, M.; Simmons, B.; Wu, M.; Betts, N.M.; Du, M.; Lyons, T.J. Strawberries decrease atherosclerotic markers in subjects with metabolic syndrome. *Nutr. Res.* **2010**, *30*, 462–469. [CrossRef] [PubMed]

56. Basu, A.; Wilkinson, M.; Penugonda, K.; Simmons, B.; Betts, N.M.; Lyons, T.J. Freeze-dried strawberry powder improves lipid profile and lipid peroxidation in women with metabolic syndrome: Baseline and post intervention effects. *Nutr. J.* **2009**, *8*, 43. [CrossRef] [PubMed]

57. Jenkins, D.J.; Nguyen, T.H.; Kendall, C.W.; Faulkner, D.A.; Bashyam, B.; Kim, I.J.; Ireland, C.; Patel, D.; Vidgen, E.; Josse, A.R.; et al. The effect of strawberries in a cholesterol-lowering dietary portfolio. *Metabolism* **2008**, *57*, 1636–1644. [CrossRef] [PubMed]

58. Soltani, R.; Hakimi, M.; Asgary, S.; Ghanadian, S.M.; Keshvari, M.; Sarrafzadegan, N. Evaluation of the effects of *Vaccinium arctostaphylos* L. fruit extract on serum lipids and hs-CRP levels and oxidative stress in adult patients with hyperlipidemia: A randomized, double-blind, placebo-controlled clinical trial. *Evid. Based Complement. Altern. Med.* **2014**, *2014*, 217451. [CrossRef] [PubMed]

59. Tjelle, T.E.; Holtung, L.; Bøhn, S.K.; Aaby, K.; Thoresen, M.; Wiik, S.Å.; Paur, I.; Karlsen, A.S.; Retterstøl, K.; Iversen, P.O.; et al. Polyphenol-rich juices reduce blood pressure measures in a randomized controlled trial in high normal and hypertensive volunteers. *Br. J. Nutr.* **2015**, *114*, 1054–1063. [CrossRef] [PubMed]

60. Puupponen-Pimiä, R.; Seppänen-Laakso, T.; Kankainen, M.; Maukonen, J.; Törrönen, R.; Kolehmainen, M.; Leppänen, T.; Moilanen, E.; Nohynek, L.; Aura, A.M.; et al. Effects of ellagitannin-rich berries on blood lipids, gut microbiota, and urolithin production in human subjects with symptoms of metabolic syndrome. *Mol. Nutr. Food Res.* **2013**, *57*, 2258–2263. [CrossRef] [PubMed]

61. Lehtonen, H.M.; Suomela, J.P.; Tahvonen, R.; Vaarno, J.; Venojärvi, M.; Viikari, J.; Kallio, H. Berry meals and risk factors associated with metabolic syndrome. *Eur. J. Clin. Nutr.* **2010**, *64*, 614–621. [CrossRef] [PubMed]

62. Zaho, Y. *Berry Fruit: Value-Added Products for Health Promotion*; CRC Press: Boca Raton, FL, USA, 2007.

63. Beattie, J.; Crozier, A.; Duthie, G.G. Potential health benefits of berries. *Curr. Nutr. Food Sci.* **2005**, *1*, 71–86. [CrossRef]

64. Beltrán-Sánchez, H.; Harhay, M.O.; Harhay, M.M.; McElligott, S. Prevalence and trends of Metabolic Syndrome in the adult US population, 1999–2010. *J. Am. Coll. Cardiol.* **2013**, *62*, 697–703. [CrossRef] [PubMed]

antioxidants

MDPI

Review

Recent Progress in Anti-Obesity and Anti-Diabetes Effect of Berries

Takanori Tsuda

College of Bioscience and Biotechnology, Chubu University, Kasugai, Aichi 487-8501, Japan;
tsudat@isc.chubu.ac.jp; Tel./Fax: +81-568-51-9659

Academic Editor: Dorothy Klimis-Zacas
Received: 29 January 2016; Accepted: 29 March 2016; Published: 6 April 2016

Abstract: Berries are rich in polyphenols such as anthocyanins. Various favorable functions of berries cannot be explained by their anti-oxidant properties, and thus, berries are now receiving great interest as food ingredients with "beyond antioxidant" functions. In this review, we discuss the potential health benefits of anthocyanin-rich berries, with a focus on prevention and treatment of obesity and diabetes. To better understand the physiological functionality of berries, the exact molecular mechanism of their anti-obesity and anti-diabetes effect should be clarified. Additionally, the relationship of metabolites and degradation products with health benefits derived from anthocyanins needs to be elucidated. The preventive effects of berries and anthocyanin-containing foods on the metabolic syndrome are not always supported by findings of interventional studies in humans, and thus further studies are necessary. Use of standardized diets and conditions by all research groups may address this problem. Berries are tasty foods that are easy to consume, and thus, investigating their health benefits is critical for health promotion and disease prevention.

Keywords: berries; anthocyanins; obesity; diabetes; degradation products; metabolites; bioavailability

1. Introduction

Berries are rich in polyphenols such as anthocyanins. Anthocyanins, belonging to the group flavonoids, are plant pigments that may appear red or purple, and are present in the form of glycosides [1,2]. Figure 1 shows the structures of the aglycone moieties of major anthocyanins [2]. The types and content of anthocyanins differ among varieties of berries, and depend on cultivation conditions and the timing of the harvest. The anthocyanin composition of some common berries is shown in Table 1 [3]. The components of berries are known for their strong anti-oxidant properties. It has been documented that anthocyanins have other health-related functions, such as improvement of visual and vascular function [4–7], anti-arteriosclerosis [8], anti-cancer [9,10], anti-obesity [11], anti-diabetes [12], and brain function-enhancing properties [13,14]. These favorable functions of berries cannot explained by their anti-oxidant properties alone, and thus, berries are now receiving great interest as food ingredients with "beyond antioxidant" functions. Also, the beneficial effects of berry component metabolites (e.g., anthocyanins) are becoming known.

Among a wide array of favorable functions of berries, their preventive and normalizing effect against obesity and diabetes will be discussed in this review, which will also outline findings regarding metabolism and absorption related to their anti-obesity and anti-diabetic effects. Furthermore, the latest interventional studies on the Metabolic Syndrome in humans are introduced. Finally, we summarize the future research needs regarding studies on berries' favorable health benefits.

Table 1. Anthocyanin Content in Fresh Berries (Reprinted with permission from [3]. Copyright 2008 American Chemical Society).

Berries	Anthocyanins	mg/g Extract [a]
Bilberry	cyanidin 3-galactoside	3.70
	cyanidin 3-glucoside	4.05
	cyanidin 3-arabinoside	2.54
	delphinidin 3-galactoside	4.58
	delphinidin 3-glucoside	4.73
	delphinidin 3-arabinoside	3.53
	peonidin 3-galactoside	0.46
	petunidin 3-halactoside	1.52
	petunidin 3-glucoside	2.94
	petunidin 3-arabinoside	0.84
	malvidin 3-arabinoside	0.81
	peonidin 3-glucoside/malvidin 3-galactoside	3.48
	peonidin 3-arabinoside/malvidin 3-glucoside	3.62
Blackberry	cyanidin 3-glucoside	7.17
	cyanidin 3-rutinoside	0.06
	cyanidin 3-arabinoside	0.05
	cyanidin 3-xyloside	0.47
	cyanidin 3-(6-malonoyl)glucoside	0.3
	cyanidin 3-dioxaloylglucoside	2.05
Blackcurrant	cyanidin 3-glucoside	1.1
	cyanidin 3-rutinoside	7.08
	delphinidin 3-glucoside	2.94
	delphinidin 3-rutinoside	9.79
	peonidin 3-rutinoside	0.11
	petunidin 3-rutinoside	0.18
Blueberry	cyanidin 3-galactoside	0.28
	cyanidin 3-glucoside	0.04
	cyanidin 3-arabinoside	0.12
	delphinidin 3-galactoside	1.37
	delphinisin 3-glucoside	0.13
	delphinidin 3-arabinoside	0.74
	peonidin 3-galactoside	0.15
	petunidin 3-galactoside	1.07
	petunidin 3-glucoside	0.11
	petunidin 3-arabinoside	0.46
	marlidin 3-arabinoside	1.75
	peonidin 3-glucoside/malvidin 3-galactoside	3.65
	peonidin 3-arabinoside/malvidin 3-glucoside	0.43
Strawberry	cyanidin 3-glucoside	0.09
	pelargonidin 3-glucoside	5.07

[a] Values are expressed as mean of triplicate analyses for each sample.

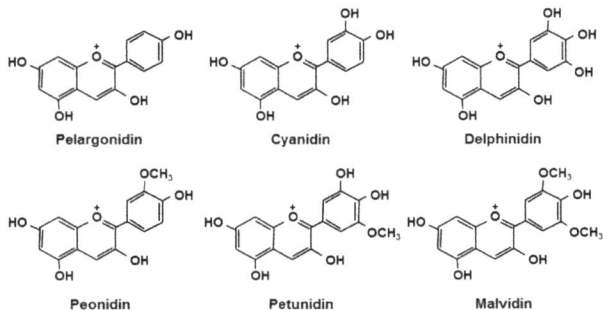

Figure 1. Chemical structures of anthocyanidins [15].

2. Preventive and Normalizing Effect on Obesity and Diabetes

2.1. Anti-Obesity

The first report that demonstrated the preventive properties of anthocyanins against body fat accumulation was published by our group in 2003 [11]. Briefly, in C57BL/6J mice, a cyanidin 3-glucoside supplemented diet (C3G; 2 g/kg) significantly suppressed body fat accumulation induced by a high-fat diet (60% fat), and this was attributed to a reduction in lipid synthesis in the liver and white adipose tissue [11,16]. Anthocyanins act on adipose tissue, inducing changes, such as those in the expression levels of adipocytokines. We have reported that C3G or its aglycones induce upregulation of adiponectin, which enhances insulin sensitivity, in isolated rat and human adipocytes [17,18], but these events were not observed *in vivo*. Prior et al. reported that feeding a high-fat diet (45% fat) supplemented with anthocyanins extracted from blueberries significantly suppressed increases in body weight and body fat accumulation in C57BL/6 mice, while intake of lyophilized wild blueberry powder (WBP) did not demonstrate the same effect but induced body fat accumulation [19]. In a separate study, the same group reported that ingestion of blueberry juice did not significantly reduce the body weight gain and the weight of white adipose tissue (epididymal and retroperitoneal fat) in mice fed a high fat diet (45% of energy fat) [20]. A different group also reported similar findings [21]. On the other hand, Seymour et al. reported that supplementation of a high-fat diet (45% fat) with 2% WBP reduced the weight of intraperitoneal fat and increased the activity of the peroxisome proliferator-activated receptor (PPAR) in white adipose tissue and skeletal muscle in Zucker fatty rats [22]. Further, Vendrame *et al.* of the University of Maine reported that 8 weeks of feeding a diet supplemented with 8% WBP significantly increased blood adiponectin levels, and decreased levels of inflammation markers in white adipose tissue [23], and ameliorated dyslipidemia [24], but did not influence fasting blood glucose and insulin levels [25] in obese Zucker rats.

The effects of other types of berries have been tested. Ingestion of black raspberries did not significantly suppress body fat accumulation and weight gain in mice fed a high-fat diet (60% fat) [26–28]. On the other hand, ingestion of an aqueous extraction of mulberries suppressed increases in body weight [29]. Intake of tart cherry power significantly reduced body weight gain and the amount of retroperitoneal fat, suppressed upregulation of obesity-related inflammatory cytokines (IL-6 and TNF-α), and increased mRNA levels of PPARα and PPARγ in Zucker fatty rats [30]. In rats fed a high-fructose diet, intake of a chokeberry extract significantly suppressed increases in the weight of epididymal fat and blood glucose level, and at the same time, it significantly increased plasma adiponectin level and decreased plasma TNF-α and IL-6 levels [31]. Taken together, the anti-obesity effect of berries is controversial and findings are not consistent among studies. Use of standardized diets and conditions in all research groups may address this problem.

2.2. Anti-Diabetes

The author's group reported that intake of purified anthocyanin (C3G) [16,32], bilberry anthocyanin extract (BBE) containing a variety of anthocyanins [12], and black soybean components (C3G and procyanidin) [33] significantly ameliorate hyperglycemia and insulin sensitivity in type 2 diabetic mice. It was reported by a different group that C3G and its metabolite protocatechuic acid caused activation of PPARγ, and also induced upregulation of Glut 4 and adiponectin in human adipocytes [34]. However, we demonstrated that C3G does not serve as a ligand of PPARγ, and did not observe C3G-induced upregulation of adiponectin *in vivo* [16,32]. Thus, it cannot be concluded that the effect of C3G against diabetes is due to activation of PPARγ-ligand or upregulation of adiponectin.

BBE activates AMP-activated protein kinase (AMPK) in the white adipose tissue and skeletal muscle. This activation induces upregulation of glucose transporter 4 and enhancement of glucose uptake and utilization in these tissues. In the liver, dietary BBE suppresses gluconeogenesis (downregulation of glucose-6-phosphatase and phosphoenol pyruvate carboxykinase) via AMPK activation, which ameliorates hyperglycemia in type 2 diabetic mice. Meanwhile, in lipid metabolism, the AMPK activation induces phosphorylation of acetylCoA carboxylase and upregulation of PPARα, acylCoA oxidase , and carnitine palmitoyltransferase-1A gene expression in the liver. These changes lead to reductions in lipid content and enhance insulin sensitivity via reduction of lipotoxicity (Figure 2) [12].

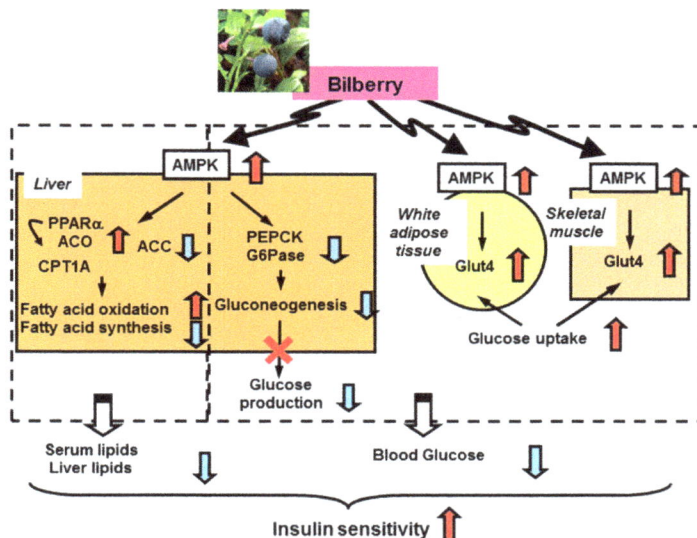

Figure 2. Proposed mechanism for amelioration of hyperglycemia and insulin sensitivity by dietary BBE [12]. ACC, acetylCoA carboxylase; ACO, acylCoA oxidase; CPT1A, carnitine palmitoyltransferase-1A; G6Pase, glucose-6-phosphatase; Glut4, glucose transporter 4; PEPCK, phosphoenol pyruvate carboxykinase.

Prevention of diabetes is an important task in the elderly and menopausal women, and ingestion of blueberries may be effective in its prevention. A research group at Louisiana State University reported that feeding a diet supplemented with 4% blueberry power for 12 weeks improved glucose intolerance and a fatty liver in post-menopausal mice [35].

Regarding preventive and suppressive effects of anthocyanins against diabetes, our group recently discovered that anthocyanins induce secretion of glucagon-like peptide-1 (GLP-1), one of the incretins. "Incretins" is a general term for a group of peptide hormones that are secreted from the gastrointestinal tract upon food ingestion and act on pancreatic βcells, thereby inducing insulin secretion in a blood glucose concentration-dependent manner. There are two known incretins, one of which is GLP-1. Because enhancement of the action of GLP-1 is effective in prevention and treatment of type 2 diabetes, inhibitors of GLP-1 degradation and degradation-resistant GLP-1 receptor agonists are used for therapeutic purposes [36,37]. Use of food-derived factors to enhance secretion of endogenous GLP-1, thereby increasing blood GLP-1 levels, would serve as a new strategy [38]. There are several molecular species of anthocyanins. Their individual effects are still unknown, but their preventive and suppressive effect against diabetes may involve the action of GLP-1. Thus, we examined the inhibitory effect of each molecular species of anthocyanins on GLP-1 secretion, and as a result, discovered delphinidin 3-rutinoside (D3R) [15]. Furthermore, we documented that such secretion-inducing effect is mediated by G-protein coupled receptors (GPR40 or 120) and the downstream calcium-calmodulin dependent protein kinase II pathway (Figure 3) [15]. These findings indicate that anthocyanins, without being absorbed, can directly act within the intestine and exert health-related effects.

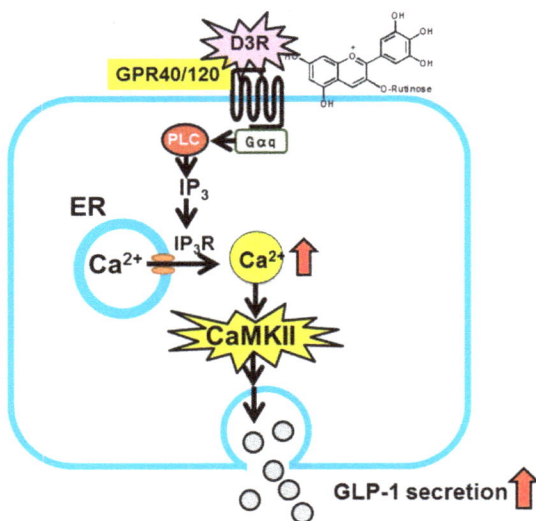

Figure 3. Proposed mechanism for stimulation of GLP-1 secretion by D3R in intestinal L-cells [15]. D3R activates G protein-coupled receptor (GPR), e.g., GPR40/120, on the L-cell surface. Activation induces IP3R-mediated release of intracellular Ca^{2+} from the endoplasmic reticulum. The elevation of cytosolic Ca^{2+} stimulates phosphorylation of Ca^{2+}/calmodulin-dependent kinaseII (CaMKII). CaMKII activation leads to an increase in GLP-1 secretion from intestinal L-cells.

3. Health Benefits of Berries and Involvement of Their Metabolites

In 1999, we reported that protocatechuic acid is a metabolite of C3G [39]. Recently, phenolic acids (protocatechuic acid, syringic acid, vanillic acid, phloroglucinol aldehyde, phloroglucinol acid and gallic acid), which are degradation products or metabolites of anthocyanins, are of great interest in relation to health benefits of berries [40–47]. These phenolic acids were detected as metabolites in humans [48]

It is believed that the bioavailability of anthocyanins is very low (about 0.1%). Also, their chemical structures indicate that they are prone to degradation, leading to the question why they exert favorable health benefits. A British group led by Kay reported a study that examined metabolism and absorption of ^{13}C-labelled C3G in humans [49]. Briefly, 500 mg ^{13}C-labelled C3G was ingested, and its excretion into the blood, urine, stool and expired air was monitored over a period of 48 h. It was shown that ^{13}C was excreted even 24–48 h after ingestion and the detected conjugates and metabolites were diverse, and that the calculated bioavailability was $\geq 12.38\% \pm 1.38\%$. In addition to conjugates of C3G and its aglycone (cyanidin), the following degradation products and metabolites were detected in the same study: protocatechuic acid and its glucuronate conjugate or sulfate conjugate; vanillic acid and its derivatives and conjugates; phenylacetic acids (3,4-dihydroxyphenyl acetic acid and 4-hydroxyphenylacetic acid; phenylpropenoic acids (caffeic acid and ferulic acid); and hippuric acid. It is likely that C3G undergoes chemical degradation in a considerably complicated metabolic processes, influenced by enteric bacteria, and is re-absorbed. The same group prepared various C3G metabolites, and reported that these metabolites at physiological concentrations suppress inflammation in human vascular endothelial cells [50,51].

A British group led by Spencer also reported that a wide range of phenolic acids in plasma, most likely to be degradation products and metabolites of anthocyanins, were present after ingestion of a blueberry drink [52]. Some of these compounds were detected in blood and peaked only 1 h after ingestion, while others were detected several hours after ingestion.

A Canadian group led by Kalt examined anthocyanin metabolites after ingestion of 250 mL blueberry juice in humans, and found that metabolites were excreted into urine over a period of 5 days [53]. This suggests that these metabolites circulate within the enterohepatic loop, and remain in the body for a long period. To confirm this, a tracer study conducted by a British group [49] can provide us with some important information. Concretely, examination of metabolism and absorption of various ^{13}C-labelled types of anthocyanin species may help confirm circulation of the metabolites within the enterohepatic loop.

A study by a group at North Carolina State University investigated whether degradation of anthocyanins by enteric bacteria and resulting products, namely phenol acids, are responsible for the health benefit of anthocyanins. It was shown that feeding mice a diet containing 1% black currant powdered extract (32% anthocyanins) for 8 weeks suppressed weight gain and improved glucose metabolism. However, these effects were observed in mice with normal gut microbiome, but not in mice with intestinal flora altered by antibiotics [54]. It is of great interest that the study showed that metabolites generated by enteric bacteria from berry components are involved in exerting health-related effects.

Taken together, the link between functional doses of anthocyanin or berry intake and metabolite concentrations may be explored to evaluate the health benefits of berries (Figure 4).

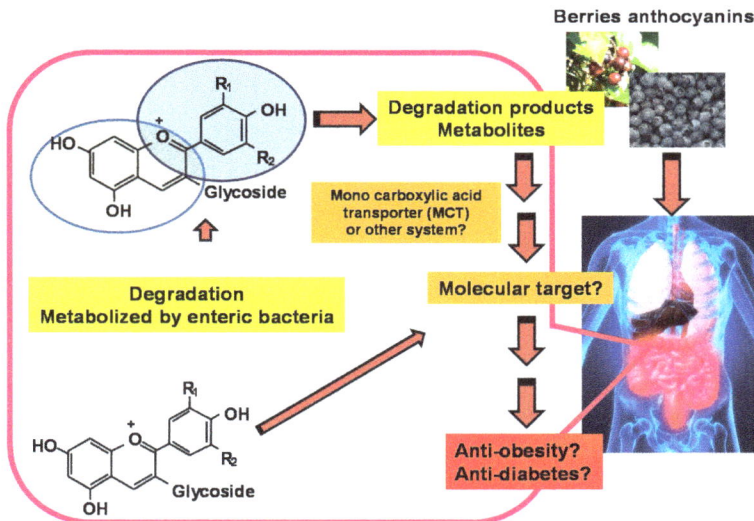

Figure 4. Degradation products or metabolites derived from berry anthocyanins can have an impact on health.

4. Interventional Studies in Humans

In this section, recent intervention studies of berries in humans are introduced. Obesity and diabetes are closely linked to cardiovascular disease. Therefore, some studies shown here include the role of berries in cardiovascular disease prevention.

Basu *et al.*, conducted a randomized study wherein 48 obese men and women (mean body mass index (BMI), 37.8) ingested lyophilized blueberries containing anthocyanins (742 mg) for 8 weeks. It was shown that blood pressure was significantly improved and oxidized low-density lipoprotein (LDL)-cholesterol levels were decreased, while the blood glucose levels, body weights and waist circumferences were not improved in the group who ingested blueberries [55]. On the other hand, a US research group conducted a randomized double-blind comparison wherein improvement of insulin sensitivity was demonstrated in 32 men and women who ingested blueberry powder (669 mg/day anthocyanins) for 6 weeks [56]. Furthermore, a research team involving US, British and Singaporean groups jointly conducted an epidemiological study, and reported significant decreases in the risk of type 2 diabetes in the group that ingested a large amount of anthocyanins [57]. A British group also reported the interrelation between anthocyanin intake and decreases in levels of blood insulin and inflammatory markers [58]. A US group reported that high anthocyanin intake significantly reduced the risk of myocardial infarction in 93,600 young and middle-aged women [59]. On the other hand, ingestion of elderberries (500 mg/day, 12 weeks) did not improve the profile of cardiovascular disease markers in 52 post-menopausal women [60]. Also, a randomized double-blind comparison found that ingestion of purple carrots (daily intake of anthocyanins and phenolic acids, 118.5 mg and 259.2 mg, respectively) for 4 weeks did not influence the body weight, LDL-cholesterol level and blood pressure, but it significantly reduced the high-density lipoprotein (HDL)-cholesterol level in 16 obese men (mean BMI, 32.8) [61]. In summary, the preventive effects of berries and anthocyanin-containing foods on the Metabolic Syndrome are currently not always supported by findings of interventional studies in humans, and thus further studies are necessary.

5. Future Research Needs and Prospects

Finally, we present the research needs on the beneficial effects of berries and anthocyanins.

First, in studies on the suppressive/normalizing effects of anthocyanins on obesity and diabetes, interrelation between chemical structures of anthocyanin molecules and their various beneficial effects remain largely unclear. On the other hand, simultaneous consumption of various types of anthocyanins may be more beneficial in some cases. Thus, it is crucial to elucidate which type of anthocyanin species or compositions of the mixture are most beneficial for their health benefits.

Second, the relationships of metabolites and degradation products derived from anthocyanins with health benefits need to be elucidated. The following questions should be answered: whether health-related effects of berries can be explained solely by metabolites and degradation products of anthocyanins; what types and quantities of metabolites and degradation products of anthocyanins are necessary for them to exert their beneficial effects; whether differences in intestinal flora influence the beneficial effects of berries; and conversely, whether ingestion of berries affects intestinal flora.

Third, it is necessary to determine whether the effects caused by anthocyanins solely explain the health benefits of berries, or the co-presence of other berry components is essential for their beneficial effects.

Lastly, health benefits that may result from berries have not yet been fully studied in humans, and findings are not consistent among studies. Use of standardized diets and conditions in all research groups may address this problem.

Berries are tasty foods that are easy to consume, and thus, investigating their health benefits is critical for health promotion and disease prevention.

Acknowledgments: This review was supported in part by Grants-in-Aid for Scientific Research (No. 26450168) from the Japan Society for Promotion of Science.

Conflicts of Interest: The author has declared no conflict of interest.

References

1. Harborne, J.B.; Grayer, R.J. *The Flavonoids*; Chapman and Hall: London, UK, 1988; pp. 1–20.
2. Tsuda, T. Dietary anthocyanin-rich plants: Biochemical basis and recent progress in health benefits studies. *Mol. Nutr. Food Res.* **2012**, *56*, 159–170. [CrossRef] [PubMed]
3. Ogawa, K.; Sakakibara, H.; Iwata, R.; Ishii, T.; Sato, T.; Goda, T.; Shimoi, K.; Kumazawa, S. Anthocyanin composition and antioxidant activity of the crowberry (*Empetrum nigrum*) and other berries. *J. Agric. Food Chem.* **2008**, *56*, 4457–4462. [CrossRef] [PubMed]

4. Nakaishi, H.; Matsumoto, H.; Tominaga, S.; Hirayama., M. Effects of blackcurrant anthocyanosides intake on dark adaptation and VDT work-induced transient refractive alternation in healthy humans. *Altern. Med. Rev.* **2000**, *5*, 553–562. [PubMed]

5. Iida, H.; Nakamura, Y.; Matsumoto, H.; Takeuchi, Y.; Harano, S.; Ishihara, M.; Katsumi, O. Effect of black-currant extract on negative lens-induced ocular growth in chicks. *Ophthalmic. Res.* **2010**, *44*, 242–250. [CrossRef] [PubMed]

6. Rissanen, T.H.; Voutilainen, S.; Virtanen, J.K.; Venho, B.; Vanharanta, M.; Mursu, J.; Salonen, J.T. Low intake of fruits, berries and vegetables is associated with excess mortality in men: the Kuopio Ischaemic Heart Disease Risk Factor (KIHD) Study. *J. Nutr.* **2003**, *133*, 199–204. [PubMed]

7. Erlund, I.; Koli, R.; Alfthan, G.; Marniemi, J.; Puukka, P.; Mustonen, P.; Mattila, P.; Jula, A. Favorable effects of berry consumption on platelet function, blood pressure, and HDL cholesterol. *Am. J. Clin. Nutr.* **2008**, *87*, 323–331. [PubMed]

8. Ellingsen, I.; Hjerkinn, E.M.; Seljeflot, I.; Arnesen, H.; Tonstad, S. Consumption of fruit and berries is inversely associated with carotid atherosclerosis in elderly men. *Br. J. Nutr.* **2008**, *99*, 674–681. [CrossRef] [PubMed]

9. Thomasset, S.; Teller, N.; Cai, H.; Marko, D.; Berry, D.P.; Steward, W.P.; Gescher, A.J. Do anthocyanins and anthocyanidins, cancer chemopreventive pigments in the diet, merit development as potential drugs? *Cancer Chemother. Pharmacol.* **2009**, *64*, 201–211. [CrossRef] [PubMed]

10. Prasad, S.; Phromnoi, K.; Yadav, V.R.; Chaturvedi, M.M.; Aggarwal, B.B. Targeting inflammatory pathways by flavonoids for prevention and treatment of cancer. *Planta Med.* **2010**, *76*, 1044–1063. [CrossRef] [PubMed]

11. Tsuda, T.; Horio, F.; Uchida, K.; Aoki, H.; Osawa, T. Dietary cyanidin 3-O-beta-D-glucoside-rich purple corn color prevents obesity and ameliorates hyperglycemia in mice. *J. Nutr.* **2003**, *133*, 2125–2130. [PubMed]

12. Takikawa, M.; Inoue, S.; Horio, F.; Tsuda, T. Dietary anthocyanin-rich bilberry extract ameliorates hyperglycemia and insulin sensitivity via activation of AMP-activated protein kinase in diabetic mice. *J. Nutr.* **2010**, *140*, 527–533. [CrossRef] [PubMed]

13. Krikorian, R.; Shidler, M.D.; Nash, T.A.; Kalt, W.; Vinqvist-Tymchuk, M.R.; Shukitt-Hale, B.; Joseph, J.A. Blueberry supplementation improves memory in older adults. *J. Agric. Food Chem.* **2010**, *58*, 3996–4000. [CrossRef] [PubMed]

14. Spencer, J.P. The impact of fruit flavonoids on memory and cognition. *Br. J. Nutr.* **2010**, *104*, S40–S47. [CrossRef] [PubMed]

15. Kato, M.; Tani, T.; Terahara, N.; Tsuda, T. The anthocyanin delphinidin 3-rutinoside stimulates glucagon-like peptide-1 secretion in murine GLUTag cell line via the Ca^{2+}/calmodulin-dependent kinase II pathway. *PLoS ONE* **2015**, *10*, e0126157.

16. Tsuda, T. Regulation of adipocyte function by anthocyanins: Possibility of preventing the metabolic syndrome. *J. Agric. Food Chem.* **2008**, *56*, 642–646. [CrossRef] [PubMed]

17. Tsuda, T.; Ueno, Y.; Aoki, H.; Koda, T.; Horio, F.; Takahashi, N.; Kawada, T.; Osawa, T. Anthocyanin enhances adipocytokine secretion and adipocyte-specific gene expression in isolated rat adipocytes. *Biochem. Biophys. Res. Commun.* **2004**, *316*, 149–157. [CrossRef] [PubMed]

18. Tsuda, T.; Ueno, Y.; Yoshikawa, T.; Kojo, H.; Osawa, T. Microarray profiling of gene expression in human adipocytes in response to anthocyanins. *Biochem. Pharmacol.* **2006**, *71*, 1184–1197. [CrossRef] [PubMed]

19. Prior, R.L.; Wu, X.; Gu, L.; Hager, T.J.; Hager, A.; Howard, L.R. Whole berries versus berry anthocyanins: interactions with dietary fat levels in the C57BL/6J mouse model of obesity. *J. Agric. Food Chem.* **2008**, *56*, 647–653. [CrossRef] [PubMed]

20. Prior, R.L.; Wilkes, S.E.; Rogers, T.R.; Khanal, R.C.; Wu, X.; Howard, L.R. Purified blueberry anthocyanins and blueberry juice alter development of obesity in mice fed an obesogenic high-fat diet. *J. Agric. Food Chem.* **2010**, *58*, 3970–3976. [CrossRef] [PubMed]

21. DeFuria, J.; Bennett, G.; Strissel, K.J.; Perfield, J.W., 2nd; Milbury, P.E.; Greenberg, A.S.; Obin, M.S. Dietary blueberry attenuates whole-body insulin resistance in high fat-fed mice by reducing adipocyte death and its inflammatory sequelae. *J. Nutr.* **2009**, *139*, 1510–1516. [CrossRef] [PubMed]

22. Seymour, E.M.; Tanone, I.I.; Urcuyo-Llanes, D.E.; Lewis, S.K.; Kirakosyan, A.; Kondoleon, M.G.; Kaufman, P.B.; Bolling, S.F. Blueberry intake alters skeletal muscle and adipose tissue peroxisome proliferator-activated receptor activity and reduces insulin resistance in obese rats. *J. Med. Food.* **2011**, *14*, 1511–1518. [CrossRef] [PubMed]

23. Vendrame, S.; Daugherty, A.; Kristo, A.S.; Riso, P.; Klimis-Zacas, D. Wild blueberry (*Vaccinium angustifolium*) consumption improves inflammatory status in the obese Zucker rat model of the metabolic syndrome. *J. Nutr. Biochem.* **2013**, *24*, 1508–1512. [CrossRef] [PubMed]

24. Vendrame, S.; Daugherty, A.; Kristo, A.S.; Klimis-Zacas, D. Wild blueberry (*Vaccinium angustifolium*)-enriched diet improves dyslipidaemia and modulates the expression of genes related to lipid metabolism in obese Zucker rats. *Br. J. Nutr.* **2014**, *111*, 194–200.

25. Vendrame, S.; Zhao, A.; Merrow, T.; Klimis-Zacas, D. The effects of wild blueberry consumption on plasma markers and gene expression related to glucose metabolism in the obese Zucker rat. *J. Med. Food.* **2015**, *18*, 619–624. [CrossRef] [PubMed]

26. Prior, R.L.; Wu, X.; Gu, L.; Hager, T.; Wilkes, S.; Howard, L. Purified berry anthocyanins but not whole berries normalize lipid parameters in mice fed an obesogenic high fat diet. *Mol. Nutr. Food Res.* **2009**, *53*, 1406–1418. [CrossRef] [PubMed]

27. Prior, R.L.; Wilkes, S.; Rogers, T.; Khanal, R.C.; Wu, X.; Howard, L.R. Dietary black raspberry anthocyanins do not alter development of obesity in mice fed an obesogenic high-fat diet. *J. Agric. Food Chem.* **2010**, *58*, 3977–3983. [CrossRef] [PubMed]

28. Kaume, L.; Howard, L.R.; Devareddy, L. The Blackberry Fruit: A Review on Its Composition and Chemistry, Metabolism and Bioavailability, and Health Benefits. *J. Agric. Food Chem.* **2012**, *60*, 5716–5727. [CrossRef] [PubMed]

29. Peng, C.H.; Liu, L.K.; Chuang, C.M.; Chyau, C.C.; Huang, C.N.; Wang, C.J. Mulberry water extracts possess an anti-obesity effect and ability to inhibit hepatic lipogenesis and promote lipolysis. *J. Agric. Food Chem.* **2011**, *59*, 2663–2671. [CrossRef] [PubMed]

30. Seymour, E.M.; Lewis, S.K.; Urcuyo-Llanes, D.E.; Tanone, I.I.; Kirakosyan, A.; Kaufman, P.B.; Bolling, S.F. Regular tart cherry intake alters abdominal adiposity, adipose gene transcription, and inflammation in obesity-prone rats fed a high fat diet. *J. Med. Food.* **2009**, *12*, 935–942. [CrossRef] [PubMed]

31. Qin, B.; Anderson, R.A. An extract of chokeberry attenuates weight gain and modulates insulin, adipogenic and inflammatory signalling pathways in epididymal adipose tissue of rats fed a fructose-rich diet. *Br. J. Nutr.* **2011**, *108*, 581–587. [CrossRef] [PubMed]

32. Sasaki, R.; Nishimura, N.; Hoshino, H.; Isa, Y.; Kadowaki, M.; Ichi, T.; Tanaka, A.; Nishiumi, S.; Fukuda, I.; Ashida, H.; *et al.* Cyanidin 3-glucoside ameliorates hyperglycemia and insulin sensitivity due to downregulation of retinol binding protein 4 expression in diabetic mice. *Biochem. Pharmacol.* **2007**, *74*, 1619–1627. [CrossRef] [PubMed]

33. Kurimoto, Y.; Shibayama, S.; Inoue, S.; Soga, M.; Takikawa, M.; Ito, C.; Nanba, F.; Yoshida, T.; Yamashita, Y.; Ashida, H.; Tsuda, T. Black soybean seed coat extract ameliorates hyperglycemia and insulin sensitivity via the activation of AMP-activated protein kinase in diabetic mice. *J. Agric. Food Chem.* **2013**, *61*, 5558–5564. [CrossRef] [PubMed]

34. Scazzocchio, B.; Varì, R.; Filesi, C.; D'Archivio, M.; Santangelo, C.; Giovannini, C.; Iacovelli, A.; Silecchia, G.; Li Volti, G.; Galvano, F.; *et al.* Cyanidin-3-*O*-β-glucoside and protocatechuic acid exert insulin-like effects by upregulating PPARγ activity in human omental adipocytes. *Diabetes* **2011**, *60*, 2234–2244. [CrossRef] [PubMed]

35. Elks, C.M.; Terrebonne, J.D.; Ingram, D.K.; Stephens, J.M. Blueberries improve glucose tolerance without altering body composition in obese postmenopausal mice. *Obesity* **2015**, *23*, 573–580. [CrossRef] [PubMed]

36. Herman, G.A.; Bergman, A.; Stevens, C.; Kotey, P.; Yi, B.; Zhao, P.; Dietrich, B.; Golor, G.; Schrodter, A.; Keymeulen, B.; *et al.* Effect of single oral doses of sitagliptin, a dipeptidyl peptidase-4 inhibitor, on incretin and plasma glucose levels after an oral glucose tolerance test in patients with type 2 diabetes. *J. Clin. Endocrinol. Metab.* **2006**, *91*, 4612–4619. [CrossRef] [PubMed]

37. Vilsbøll, T.; Brock, B.; Perrild, H.; Levin, K.; Lervang, H.H.; Kølendorf, K.; Krarup, T.; Schmitz, O.; Zdravkovic, M.; Le-Thi, T.; *et al.* Liraglutide, a once-daily human GLP-1 analogue, improves pancreatic B-cell function and arginine-stimulated insulin secretion during hyperglycaemia in patients with Type 2 diabetes mellitus. *Diabet. Med.* **2008**, *25*, 152–156. [CrossRef] [PubMed]

38. Tsuda, T. Possible abilities of dietary factors to prevent and treat diabetes via the stimulation of glucagon-like peptide-1 secretion. *Mol. Nutr. Food Res.* **2015**, *59*, 1264–1273. [CrossRef] [PubMed]

39. Tsuda, T.; Horio, F.; Osawa, T. Absorption and metabolism of cyanidin 3-*O*-β-D-glucoside in rats. *FEBS Lett.* **1999**, *449*, 179–182. [CrossRef]

40. Keppler, K.; Humpf, H.-U. Metabolism of anthocyanis and their phenolic degradation products by the intestinal microflora. *Bioorg. Med. Chem.* **2005**, *13*, 5195–5205. [CrossRef] [PubMed]

41. Aura, A.M.; Martin-Lopez, P.; O'Leary, K.-A.; Williamson, G.; Oksman-Caldentey, K.M.; Poutanen, K.; Santos-Buelga, C. *In vitro* metabolism of anthocyanins by human gut microflora. *Eur. J. Nutr.* **2005**, *44*, 133–142. [CrossRef] [PubMed]

42. He, J.; Magnuson, B.A.; Giusti, M.M. Analysis of anthocyanins in rat intestinal contents—Impact of anthocyanin chemical structure on fecal excretion. *J. Agric. Food Chem.* **2005**, *53*, 2859–2866. [CrossRef] [PubMed]

43. Forester, S.C.; Waterhouse, A.L. Identification of Cabernet Sauvignon Anthocyanin gut microflora metabolites. *J. Agric. Food Chem.* **2008**, *56*, 9299–9304. [CrossRef] [PubMed]

44. Forester, S.C.; Waterhouse, A.L. Gut metabolites of anthocyanins, gallic acid, 3-O-methylgallic acid, and 2,4,6-trihydroxybenzaldehyde, inhibit cell proliferation of Caco-2 cells. *J. Agric. Food Chem.* **2010**, *58*, 5320–5327. [CrossRef] [PubMed]

45. Avila, M.; Hidalgo, M.; Sanchez-Moreno, C.; Pelaez, C.; Requena, T.; de Pascual-Teresa, S. Bioconversion of anthocyanin glycosides by Bifidobacteria and Lactobacillus. *Food Res. Int.* **2009**, *42*, 1453–1461. [CrossRef]

46. Gonthier, M.P.; Cheynier, V.; Donovan, J.L.; Manach, C.; Morand, C.; Mila, I.; Lapierre, C.; Rémésy, C.; Scalbert, A. Microbial aromatic acid metabolites formed in the gut account for a major fraction of the polyphenols excreted in urine of rats fed red wine polyphenols. *J. Nutr.* **2003**, *133*, 461–467. [PubMed]

47. Borges, G.; Roowi, S.; Rouanet, J.M.; Duthie, G.G.; Lean, M.E.; Crozier, A. The bioavailability of raspberry anthocyanins and ellagitannins in rats. *Mol. Nutr. Food Res.* **2007**, *51*, 714–725. [CrossRef] [PubMed]

48. Vitaglione, P.; Donnarumma, G.; Napolitano, A.; Galvano, F.; Gallo, A.; Scalfi, L.; Fogliano, V. Protocatechuic acid is the major human metabolite of cyanidin-glucosides. *J. Nutr.* **2007**, *137*, 2043–2048. [PubMed]

49. Czank, C.; Cassidy, A.; Zhang, Q.; Morrison, D.J.; Preston, T.; Kroon, P.A.; Botting, N.P.; Kay, C.D. Human metabolism and elimination of the anthocyanin, cyanidin-3-glucoside: A ^{13}C-tracer study. *Am. J. Clin. Nutr.* **2013**, *97*, 995–1003. [CrossRef] [PubMed]

50. Amin, H.P.; Czank, C.; Raheem, S.; Zhang, Q.; Botting, N.P.; Cassidy, A.; Kay, C.D. Anthocyanins and their physiologically relevant metabolites alter the expression of IL-6 and VCAM-1 in CD40L and oxidized LDL challenged vascular endothelial cells. *Mol. Nutr. Food Res.* **2015**, *59*, 1095–1106. [CrossRef] [PubMed]

51. Edwards, M.; Czank, C.; Woodward, G.M.; Cassidy, A.; Kay, C.D. Phenolic metabolites of anthocyanins modulate mechanisms of endothelial function. *J. Agric. Food Chem.* **2015**, *63*, 2423–2431. [CrossRef] [PubMed]

52. Rodriguez-Mateos, A.; Rendeiro, C.; Bergillos-MecaIntake, T.; Tabatabaee, S.; George, T.W.; Heiss, C.; Spencer, J.P. Intake and time dependence of blueberry flavonoid–induced improvements in vascular function: a randomized, controlled, double-blind, crossover intervention study with mechanistic insights into biological activity. *Am. J. Clin. Nutr.* **2013**, *98*, 1179–1191. [CrossRef] [PubMed]

53. Kalt, W.; Liu, Y.; McDonald, J.E.; Vinqvist-Tymchuk, M.R.; Fillmore, S.A. Anthocyanin metabolites are abundant and persistent in human urine. *J. Agric. Food Chem.* **2014**, *62*, 3926–3934. [CrossRef] [PubMed]

54. Esposito, D.; Damsud, T.; Wilson, M.; Grace, M.H.; Strauch, R.; Li, X.; Lila, M.A.; Komarnytsky, S. Black currant anthocyanins attenuate weight gain and improve glucose metabolism in diet-induced obese mice with intact, but not disrupted, gut microbiome. *J. Agric. Food Chem.* **2015**, *63*, 6172–6180. [CrossRef] [PubMed]

55. Basu, A.; Du, M.; Leyva, M.J.; Sanchez, K.; Betts, N.M.; Wu, M.; Aston, C.E.; Lyons, T.J. Blueberries decrease cardiovascular risk factors in obese men and women with metabolic syndrome. *J. Nutr.* **2010**, *140*, 1582–1587. [CrossRef] [PubMed]

56. Stull, A.J.; Cash, K.C.; Johnson, W.D.; Champagne, C.M.; Cefalu, W.T. Bioactives in blueberries improve insulin sensitivity in obese, insulin-resistant men and women. *J. Nutr.* **2010**, *140*, 1764–1768. [CrossRef] [PubMed]

57. Wedick, N.M.; Pan, A.; Cassidy, A.; Rimm, E.B.; Sampson, L.; Rosner, B.; Willett, W.; Hu, F.B.; Sun, Q.; van Dam, R.M. Dietary flavonoid intakes and risk of type 2 diabetes in US men and women. *Am. J. Clin. Nutr.* **2012**, *95*, 925–933. [CrossRef] [PubMed]

58. Jennings, A.; Welch, A.A.; Spector, T.; Macgregor, A.; Cassidy, A. Intakes of anthocyanins and flavones are associated with biomarkers of insulin resistance and inflammation in women. *J. Nutr.* **2014**, *144*, 202–208. [CrossRef] [PubMed]

59. Cassidy, A.; Mukamal, K.J.; Liu, L.; Franz, M.; Eliassen, A.H.; Rimm, E.B. High anthocyanin intake is associated with a reduced risk of myocardial infarction in young and middle-aged women. *Circulation* **2013**, *127*, 188–196. [CrossRef] [PubMed]
60. Curtis, P.J.; Kroon, P.A.; Hollands, W.J.; Walls, R.; Jenkins, G.; Kay, C.D.; Cassidy, A. Cardiovascular disease risk biomarkers and liver and kidney function are not altered in postmenopausal women after ingesting an elderberry extract rich in anthocyanins for 12 weeks. *J. Nutr.* **2009**, *139*, 2266–2271. [CrossRef] [PubMed]
61. Wright, O.R.; Netzel, G.A.; Sakzewski, A.R. A randomized, double-blind, placebo-controlled trial of the effect of dried purple carrot on body mass, lipids, blood pressure, body composition, and inflammatory markers in overweight and obese adults: the QUENCH trial. *Can. J. Physiol. Pharmacol.* **2013**, *91*, 480–488. [CrossRef] [PubMed]

antioxidants

MDPI

Review

Blueberries' Impact on Insulin Resistance and Glucose Intolerance

April J. Stull

Department of Human Ecology, University of Maryland Eastern Shore, Princess Anne, MD 21853, USA; ajstull@umes.edu; Tel.: +1-410-651-6060

Academic Editor: Dorothy Klimis-Zacas
Received: 17 October 2016; Accepted: 18 November 2016; Published: 29 November 2016

Abstract: Blueberries are a rich source of polyphenols, which include anthocyanin bioactive compounds. Epidemiological evidence indicates that incorporating blueberries into the diet may lower the risk of developing type 2 diabetes (T2DM). These findings are supported by pre-clinical and clinical studies that have shown improvements in insulin resistance (i.e., increased insulin sensitivity) after obese and insulin-resistant rodents or humans consumed blueberries. Insulin resistance was assessed by homeostatic model assessment-estimated insulin resistance (HOMA-IR), insulin tolerance tests, and hyperinsulinemic-euglycemic clamps. Additionally, the improvements in glucose tolerance after blueberry consumption were assessed by glucose tolerance tests. However, firm conclusions regarding the anti-diabetic effect of blueberries cannot be drawn due to the small number of existing clinical studies. Although the current evidence is promising, more long-term, randomized, and placebo-controlled trials are needed to establish the role of blueberries in preventing or delaying T2DM.

Keywords: blueberries; bilberries; strawberries; cranberries; berries; anthocyanins; diabetes; insulin; glucose; diabetes

1. Introduction

Insulin resistance is a public health concern that can initially occur in the prediabetes stage many years before the diagnosis of type 2 diabetes mellitus (T2DM). Insulin resistance is defined as inefficient glucose uptake and utilization in peripheral tissues in response to insulin stimulation [1]. Insulin resistance in the prediabetes stage is a characteristic of glucose intolerance, which includes impaired fasting glucose (fasting plasma glucose (FPG) 100–125 mg/dL or 5.6–6.9 mmol/L) and/or impaired glucose tolerance (oral glucose tolerance test (OGTT) 2-h plasma glucose (PG) 140–199 mg/dL or 7.8–11.0 mmol/L) [2,3]. Prediabetes is a condition in which blood glucose levels are higher than normal, but not high enough to be classified as T2DM. The prediabetes stage is when corrective actions need to be implemented in order to prevent or delay the development of T2DM (FPG $\geq$ 126 mg/dL or $\geq$ 7.0 mmol/L; or OGTT 2-h PG $\geq$ 200 mg/dL or $\geq$ 11.1 mmol/L). Thirty-seven percent of adult Americans have prediabetes, which increases their risk of developing T2DM and cardiovascular disease [4]. To circumvent the health complications of T2DM and its related financial burdens, primary prevention before the disease actually occurs is warranted.

Lifestyle and dietary habits are major factors determining the development and progression of T2DM. Epidemiological studies reported that consumption of foods rich in anthocyanins, particularly from blueberries, were associated with a lower risk of T2DM and index of peripheral insulin resistance [5–7]. Blueberries belong to the genus *Vaccinium* and their health benefits may be attributable to the bioactive compounds, anthocyanins, which also have antioxidant properties [8–10]. Anthocyanins are polyphenols that belong to the flavonoid subgroup and they are the natural dark pigment color in plant foods [11]. The bioactive compounds are abundant in fruits and vegetables, such as berries, cherries, grapes, red onion, red radish, and purple potatoes [12].

Blueberries have become a popular fruit that gained the interest of the public and scientific communities due to their role in maintaining and improving health [13]. The scientific evidence supporting the anti-diabetic health benefits of blueberries is growing. Pre-clinical [14–21] and clinical [22–25] studies have found improvements in insulin resistance and glucose tolerance after blueberry consumption in obese and insulin-resistant rodents and humans. For many years, increased consumption of blueberries has been a folk remedy in Canada for treating T2DM [26]. This review will examine the effects of blueberries on insulin resistance and glucose intolerance, including evidence from dietary intervention studies that used rodents or humans with T2DM or at risk of developing the disease. An overview of mechanistic insights from cell culture and gut hormones will be explored. In addition, the review will also highlight the anti-diabetic effect of bilberries, which also belong to the genus *Vaccinium* and are known outside the United States as the "European blueberry".

2. Anti-Diabetic Effect of Blueberries

2.1. Preclinical Dietary Interventions

Increasing insulin sensitivity (i.e., improving insulin resistance and glucose tolerance) is important in preventing or improving T2DM. A number of animal studies have demonstrated the anti-diabetic effects of blueberry anthocyanins (Figure 1). Obese rodents that were diet-induced and genetically manipulated consumed a 45%–60% high fat-diet (HFD) with blueberries for 3–12 weeks and their insulin resistance (i.e., assessed using the homeostatic model assessment-estimated insulin resistance (HOMA-IR)) [18,21] and glucose tolerance (assessed using the glucose tolerance test) [15–18,20] were improved. Similar results were observed when an intraperitoneal insulin tolerance test (ITT) was used to measure insulin sensitivity. DeFuria et al. [14] found that C57BL/6 mice that consumed a 60% HFD + 4% blueberries for 8 weeks had a lower plasma glucose area under the curve (AUC) (i.e., increased insulin sensitivity) during an ITT compared with the mice fed the HFD alone. The mice on the HFD + blueberries had similar results to the 10% low-fat diet fed mice. Additionally, similar increases in insulin sensitivity (assessed by ITT) were found in diabetic KKA^y mice that consumed a bilberry diet for 5 weeks [19].

In opposition to the previous positive anti-diabetic blueberry studies, other researchers documented no influence of blueberries on insulin resistance and/or glucose tolerance in obese mice and rats [15,27–29]. Although Vendrame and Colleagues [29] did not observe any significant changes on HOMA-IR with blueberry supplementation, they did find significant biological changes in the glucose metabolism related plasma markers (hemoglobin A1c, retinol-binding protein 4, and resistin concentrations). These markers were lower in the obese Zucker rats that consumed a 8% blueberry diet for 8 weeks when compared to the rats that did not consume blueberries. Also, the gene expression related to glucose metabolism (resistin in liver and retinol-binding protein 4 in adipose tissue) was downregulated in the obese Zucker rats following blueberry intake [29].

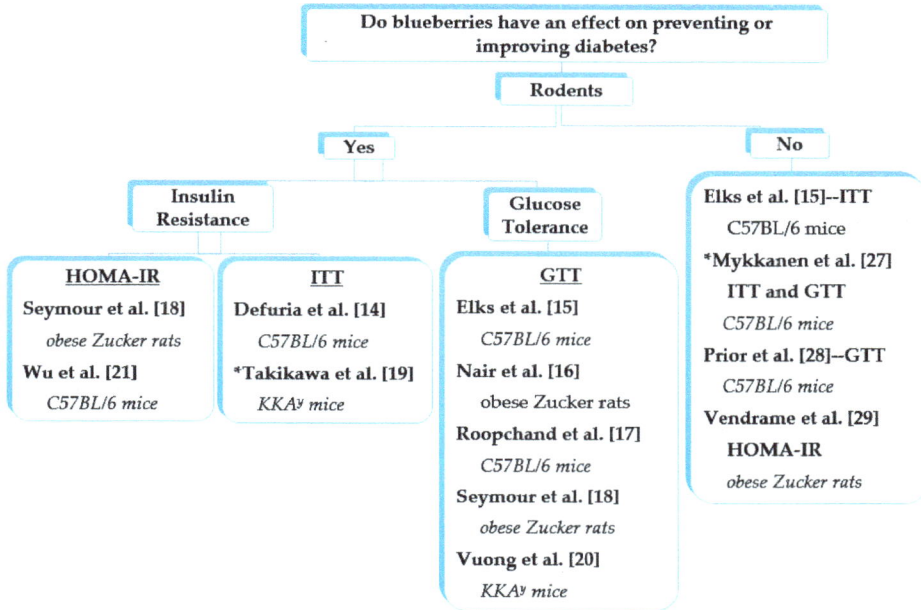

Figure 1. The effect of blueberries on preventing and improving type 2 diabetes in obese C57BL/6 mice, KKA^y mice, and Zucker rats. The rodents were fed blueberries for over 3 weeks and insulin resistance and/or glucose tolerance were assessed using HOMA-IR (homeostatic model assessment-estimated insulin resistance), ITT (insulin tolerance test), and GTT (glucose tolerance test). Seymour et al. [18], Mykkanen et al. [27], and Elks et al. [15] evaluated insulin resistance and glucose tolerance. * Studies that used bilberries.

2.2. Clinical Dietary Interventions

2.2.1. Whole Blueberries

In humans, evidence of blueberries impacting insulin resistance is sparse (Figure 2). Our lab group was the first to publish a report on the clinical impact of blueberries on whole-body insulin sensitivity in a population that was at risk for developing T2DM [10]. We found that consuming a smoothie supplemented with blueberries for 6 weeks had a greater increase in insulin sensitivity in obese and insulin-resistant adults (i.e., prediabetes) when compared to their counterparts that consumed a placebo smoothie. Insulin sensitivity was assessed by using the "gold standard" hyperinsulinemic-euglycemic clamp. Other studies, including our lab, have used less sensitive methods such as HOMA-IR [30] and frequently sampled intravenous glucose tolerance test (FSIVGTT) [31] to evaluate insulin sensitivity as a secondary measurement. Using these less sensitive methods resulted in no changes in insulin sensitivity between the blueberry and placebo groups.

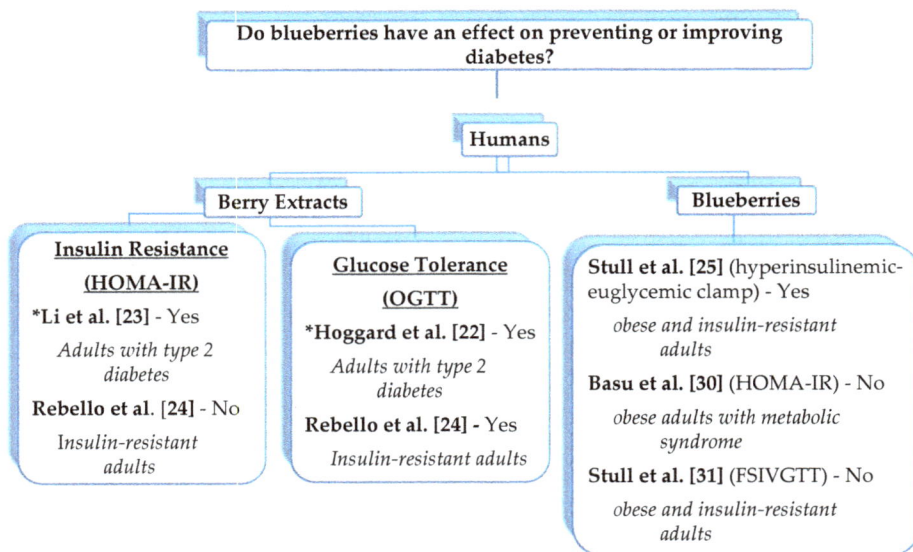

Figure 2. The effect of blueberries on preventing and improving type 2 diabetes in obese and insulin-resistant adults. Insulin resistance and/or glucose tolerance were assessed using HOMA-IR (homeostatic model assessment-estimated insulin resistance), FSIVGTT (Frequently sampled intravenous glucose tolerance test), and OGTT (oral glucose tolerance test). Rebello et al. [24], used HOMA-IR and OGTT and Stull et al. [25,31], used the clamp and FSIVGTT. * Studies that used bilberries.

2.2.2. Blueberry or Bilberry Extracts

There are clinical studies that supplemented subjects with the blueberry or bilberry extracts instead of the whole berry (Figure 2). In overweight and obese subjects, Rebello et al. [24] used a gastrointestinal microbiome modulator (GIMM) containing inulin from agave, β-glucan from oats, and polyphenols from blueberry pomace as the dietary intervention. Consuming GIMM over 4 weeks improved glucose tolerance (assessed by oral glucose tolerance test; OGTT), but no changes in insulin resistance (assessed by HOMA-IR) were observed. Li and colleagues reported further evidence supporting the anti-diabetic role of berry extracts. The subjects with T2DM that consumed capsules containing 80 mg of anthocyanins (purified from the bilberry and blackcurrent; twice daily) for 24 weeks had a significant improvement in HOMA-IR (i.e., increased insulin sensitivity) [23]. Another study with a similar population distributed a single oral capsule of either 0 g (placebo) or 0.47 g standardized bilberry extract (36% w/w anthocyanins) to the subjects with T2DM [22]. This acute crossover design study found that supplementation with the bilberry extract resulted in a lower incremental plasma glucose and insulin (assessed by OGTT) when compared to consuming the placebo.

2.3. Ingredients in the Blueberry and Placebo Drinks, Pellets, or Capsules

The blueberry and placebo drinks, pellets, or capsules differed between the human and animal studies that were reviewed in Table 1. The reviewed studies varied in the types of berries, berry extract combinations, methods of administering the treatments (whole berry vs berry extracts), and contents in the food matrix. Potential concerns with the placebo that could influence the outcome data were not having a matched macronutrient placebo that was similar to the treatment and not controlling for fiber in the placebo. Thus, fiber has been shown to positively affect glucose control [32]. Another potential problem with the placebo is the added dark dye to make it indistinguishable from the treatment intervention. The chemical structures of the dark dyes are closely related to anthocyanins and this could possibly affect the study's outcome variables. Regarding the food matrix, the smoothies in Stull et al.'s [25,31] study contained milk and yogurt and there is controversy about whether the proteins in milk interact with polyphenols and negate their antioxidant capacity and bioavailability. However, there are still discrepancies between studies [33–37]. It is important to note that the milk contained in the blueberry smoothie did not mask the beneficial effects of the blueberries on improving insulin sensitivity and endothelial function [25,31].

Table 1. Ingredients in the Blueberry Treatment and Placebo Drinks, Pellets, or Capsules.

	Study Type	BB Treatment	Placebo
Blueberries			
Lowbush (wild)			
Vuong et al. [20]	Pre-Clinical	BB juice (40 mL·kg^{-1} per day in drinking water)	water
Prior et al. [28]	Pre-Clinical	10% BB + LFD or HFD (pellets)	LFD or HFD (pellets)
Vendrame et al. [29]	Pre-Clinical	8% BB (pellets; regular diet)	pellets; regular diet
Highbush			
Defuria et al. [14]	Pre-Clinical	4% BB + HFD (pellets)	HFD (pellets)
Elks et al. [15]	Pre-Clinical	4% BB + HFD (pellets)	HFD (pellets)
Roopchand et al. [17]	Pre-Clinical	40% BB-defatted soyben flour (DSF) + HFD	HFD + DSF
Seymour et al. [18]	Pre-Clinical	2% BB + HFD (Semipurified diet)	HFD (Semipurified diet)
Stull et al. [25,31]	Clinical	22.5 g BB; 12 oz smoothie (yogurt and milk; 4 g Fiber) (twice daily)	12 oz smoothie (food color, BB flavor, and 4 g fiber) (twice daily)
Basu et al. [30]	Clinical	25g BB + 480 ml water (twice daily)	480 mL water (twice daily)
Extract			
Rebello et al. [24]	Clinical	BB ACN and polyphenols + 8.7 g fiber + 6 oz water (twice daily)	8.7 g fiber + 6 oz water (twice daily)
Unknown (Highbush or Lowbush)			
Nair et al. [16]	Pre-Clinical	2% BB + regular diet	corn + regular diet
Wu et al. [21]	Pre-Clinical	HFD + BB juice	HFD + water
Bilberries			
Takikawa et al. [19]	Pre-Clinical	27 g BB/kg + laboratory diet	laboratory diet
Mykkanen et al. [27]	Pre-Clinical	5% or 10% BB + HFD (pellets)	HFD (pellets)
Hoggard et al. [22]	Clinical	single gelatin capsule; 0.47 g of Mirtoselect® (a standardized BB extract (36 % (w/w) of anthocyanins); ~50 g fresh BB	microcrystalline cellulose in an opaque single gelatin capsule
Li et al. [23]	Clinical	80 mg BB ACN + pullulan + maltodextrin capsule (twice daily)	pullulan + maltodextrin capsule (twice daily)

Abbreviations used: BB = blueberry or bilberry, HFD = 45% or 60% kcal high fat diet, ACN = anthocyanin.

3. Prevention of Obesity-Potential Factor That May Contribute to the Anti-Diabetic Effect of Blueberries

The improved insulin sensitivity after blueberry supplementation that is exhibited in studies could possibly be due to the observed body weight and adiposity reduction in rodents. It is known that obesity is a major contributor to insulin resistance and changes in adiposity can greatly alter insulin sensitivity [38]. As seen in obesity, accumulation of lipids in tissues is a key step in the initiation and progression of insulin resistance to T2DM [38]. Increasing insulin sensitivity is important in preventing the development of T2DM. When blueberries are added to the diet, some studies have reported that obese rodents display a decrease in body weight gain and/or lipid accumulation in tissues with increased insulin sensitivity [17,20,21]. Contrarily, Prior et al. [28] observed increases in body weight gain and adiposity with blueberry consumption and this could possibly explain why blueberries did not influence insulin sensitivity in the obese mice. However, protection against obesity was observed when the obese mice were fed purified anthocyanins from blueberries [28,39]. Other researchers demonstrated increases in insulin sensitivity after blueberry consumption, but the blueberries were ineffective in reducing body weight gain and adiposity in the obese rodents [14,16,19]. Mykkanen et al. [27] observed the opposite and there were reductions in the body weight gain after bilberry supplementation in obese mice, but insulin sensitivity was not affected. In addition, Seymour et al. and colleagues [18] incorporated blueberries in the diet and the abdominal fat was reduced along with increases in insulin sensitivity in the obese Zucker rats. However, the total fat mass and body weight gain were unchanged during the 12 week study duration [18].

Body weight and fat composition have been mostly explored in animals, and to a lesser extent in humans. In clinical studies, body weight and fat composition have been explored as secondary measurements and the blueberry intake over 6–8 weeks did not change the body composition in the obese individuals [25,30,31]. Despite no changes in body weight and fat composition, Stull et al. [25] still observed an increase in insulin sensitivity after 6 weeks of blueberry consumption. Thus, it is possible that blueberries may help counteract obesity as seen in animal studies, but may not be as effective in inducing weight loss. Thus, clinical trials evaluating the anti-obesity effect of blueberries in humans is warranted with a longer study duration than 6–8 weeks.

4. Mechanisms of Action That are Related to the Anti-Diabetic Effect of Blueberries

4.1. Inhibition of Inflammatory Responses

Chronic inflammation is likely the link between obesity and insulin resistance [40,41]. Obesity is associated with macrophage infiltration into adipose tissue and the activation of the inflammatory pathway which leads to the development of insulin resistance [40,41]. The accumulation of macrophages in the adipocytes secrete proinflammatory cytokines. Previous animal studies have observed an anti-inflammatory effect of blueberries [14,27,42]. Vendrame et al. [42] reported that blueberries had an anti-inflammatory effect as evidence by a decreased expression in nuclear factor κB, interleukin-6 (IL-6), and tumor necrosis factor alpha (TNFα) in the liver and abdominal adipose tissue and decreased plasma concentrations in IL-6, TNFα, and c-reactive protein in obese Zucker rats. Insulin sensitivity was not evaluated in this particular study. Similarly, Defuria and colleagues [14] found that blueberries protected against adipocyte death, and down-regulation in gene expression indices of adipose tissue macrophage and inflammatory cytokines (TNFα and IL-10) in obese-induced mice. The researchers concluded that these changes in gene expression of inflammatory cytokines may have contributed to the increase in insulin sensitivity. Contrarily, an animal study reported increased insulin sensitivity, but no significant effect of blueberry intake on plasma inflammatory markers in obese Zucker rats [18].

In humans with hypercholesterolemia, consuming extracts from bilberry and blackcurrant anthocyanin significantly decreased the biomarker of inflammation on the vascular endothelium, soluble vascular cell adhesion molecule-1 (sVCAM-1), when compared to consuming the placebo [43]. When studies used the whole blueberry as the dietary intervention, the effects on the inflammatory response were less pronounced. Our research group [25,31] and other researchers [30,44,45] found that changes over 6–8 weeks in plasma levels of soluble intercellular adhesion molecule-1, sVCAM-1, C-reactive protein, IL-6, monocyte chemoattractant protein 1, and TNFα did not differ between the blueberry and placebo groups. Despite no changes in the inflammatory response, Stull el al. [25] still observed an increase in insulin sensitivity. Thus, in humans, a longer study duration, populations with elevated baseline inflammatory levels, and evaluation of the gene expression of the inflammatory markers may be necessary to observe an anti-inflammatory effect of blueberries on obesity and insulin resistance.

4.2. Modification of the Insulin-Dependent and Independent Cellular Pathways

There is evidence in both in vitro and in vivo models that suggest blueberries may modulate the intracellular pathways of glucose metabolism. However, there is still not a definitive answer for the cellular mechanism(s) that contribute to the anti-diabetic effect of blueberries. It is possible that there is more than one mechanism for blueberry-anthocyanins. Cell culture and animal studies have found that blueberry glucose uptake was due to activity in the insulin-dependent pathway [18,26] while other researchers have observed the activity in the insulin-independent pathway [19,46]. Contrarily, Roopchand et al. [17] found that blueberry-anthocyanins did not increase glucose uptake in L6 myotubes (i.e., skeletal muscle cells). However, these researchers did observe reduced glucose production in the H4IIE rat hepatocytes after adding blueberry anthocyanins.

Martineau et al. [26] showed that 21-h incubation of the blueberry (or fruit) extract in muscle cells enhanced glucose uptake only in the presence of insulin, which is an indication that the insulin-dependent pathway was utilized. Seymour et al. [18] reported that rats had an increase in mRNA transcripts related to glucose uptake and metabolism (e.g., insulin receptor substrate 1 (IRS 1) and glucose transporter 4 (GLUT 4)) in the skeletal muscle and retroperitoneal fat after consuming blueberries for 12 weeks. A different observation by Voung and Colleagues [46g] found the increase in glucose uptake was explained by the increased phosphorylation/activation of proteins in the insulin-independent pathway (e.g., AMP-activated protein kinase (AMPK)) in cultured muscle cells and adipocytes. Thus, the proteins in the insulin-dependent pathway (e.g., protein kinase B/AKT and extracellular signal-regulated kinase 1/2 (ERK)) were not affected by the blueberry treatments. In an in vivo study, bilberries activated AMPK in the white adipose tissue and skeletal muscle in KKA^y mice [19]. This activation induced upregulation of GLUT 4 and enhancement of glucose uptake and utilization in these tissues without using insulin. This data supported the previous evidence [46] that blueberries increased glucose uptake into the skeletal muscle cells and adipocytes via an insulin-independent mechanism.

4.3. Other Mechanisms

Anthocyanins may have various anti-diabetic effects via mechanisms other than cellular signaling proteins found in the insulin-dependent and independent pathways, such as the modification of glucagon-like peptide-1 (GLP-1), alteration of peroxisome proliferator-activated receptor (PPAR) activities, protection against glucolipotoxicity, and modification of endogenous antioxidants. It is possible that anthocyanins can act directly within the intestine and exert health related benefits. Kato et al. [47] demonstrated that delphinidin 3-rutinoside (i.e., an anthocyanin) significantly increased GLP-1 secretion in GLUTag cells via the Ca^{2+}/calmodulin-dependent kinase II pathway. GLP-1 is secreted from enteroendocrine L-cells and is one type of incretin that stimulates the glucose-dependent insulin secretion and proliferation of pancreatic β-cells. Increasing endogenous GLP-1 secretion is an alternative therapeutic approach that could possibly help treat diabetes and decrease the required medication doses [48–50]. The transcription factor PPAR was also observed and whole blueberries, isolated anthocyanins, and anthocyanin-rich extracts increased its activity [18,51–53]. PPARs are nuclear fatty acid receptors that play an important role in obesity-related metabolic diseases and PPAR agonist drugs have been used to improve insulin resistance [54]. In addition, the Chinese blueberry has been found to effectively protect β-cells against glucolipotoxicity when compared to metformin in vitro by reducing intracellular triglyceride levels, restoring intracellular insulin content, lowering basal insulin secretion, and increasing glucose-stimulated insulin secretion [55]. Glucolipotoxicity is a harmful effect of elevated glucose and fatty acid levels on pancreatic beta-cell function and survival [56]. Lastly, there are studies that demonstrated anthocyanins enhancing the endogenous antioxidant defense system. The purified anthocyanin cyanidin-3-*O*-β-glucoside increased glutathione (i.e., antioxidant) synthesis in the liver of diabetic db/db mice through upregulation of glutamate-cysteine ligase catalytic subunit expression [57]. Similar to the previous study [57], Roy and Colleagues [58] observed enhanced serum levels of superoxide dismutase and catalase after injections of pelargonidin (i.e., anthocyanidin) in the streptozotocin-induced diabetic rats.

5. Conclusions

Blueberries offer a natural "healthy package" of diverse bioactive compounds that contribute to its many health benefits. This review highlighted a multitude of in vivo and in vitro studies that demonstrated the anti-diabetic effects of blueberries and berry extracts in insulin-resistant rodent, human, and cell culture models. These beneficial effects of blueberries on insulin resistance and glucose tolerance in humans is in concordance with the animal and cell culture studies. Although there were studies that demonstrated a positive anti-diabetic effect of blueberries, this review also discussed studies with less pronounced effects. It is important to note that majority of the human studies that did not observe a positive outcome with whole blueberry supplementation used a less sensitive measurement to assess insulin sensitivity and also insulin sensitivity was a secondary measurement in the study.

The varying types of berries, berry extract combinations, the methods of administering the treatments (whole berry vs berry extracts), population studied, and the specifics of each study design bring a substantial amount of variation amongst the results in the various blueberry studies. There is a great need for more well designed, randomized, and placebo-controlled clinical trials that further explore dose responses, whole blueberries versus bioactive compounds, longevity of any health benefits, and interactions between blueberry bioactives and other foods and drugs. In addition, the cellular mechanisms are still controversial and findings are not consistent among studies. Therefore, more cellular mechanistic studies are warranted in in vivo models to elucidate the specific cellular signaling pathway involved in the improvement of insulin sensitivity after blueberry consumption. To date, there are a limited number of clinical studies that have evaluated the effect of blueberries on insulin sensitivity and more clinical trials are warranted before a definitive conclusion can be drawn about the anti-diabetic effect of blueberries.

Acknowledgments: This review was supported by the National Center for Complementary and Integrative Health (NCCIH) of the National Institutes of Health (NIH) under Award Number K01AT006975.

Conflicts of Interest: The author declare no conflict of interest.

References

1. Haffner, S.M. The insulin resistance syndrome revisited. *Diabetes Care* **1996**, *19*, 275–277. [CrossRef] [PubMed]
2. Nathan, D.M.; Davidson, M.B.; DeFronzo, R.A.; Heine, R.J.; Henry, R.R.; Pratley, R.; Zinman, B. Impaired fasting glucose and impaired glucose tolerance: Implications for care. *Diabetes Care* **2007**, *30*, 753–759. [CrossRef] [PubMed]
3. American Diabetes Association. Classification and diagnosis of diabetes. *Diabetes Care* **2016**, *39*, S13–S22.
4. Centers for Disease Control and Prevention. *National Diabetes Statistics Report: Estimates of Diabetes and Its Burden in the United States*; Department of Health and Human Services: Atlanta, GA, USA, 2014.
5. Muraki, I.; Imamura, F.; Manson, J.E.; Hu, F.B.; Willett, W.C.; van Dam, R.M.; Sun, Q. Fruit consumption and risk of type 2 diabetes: Results from three prospective longitudinal cohort studies. *BMJ* **2013**, *347*, f5001. [CrossRef] [PubMed]
6. Jennings, A.; Welch, A.A.; Spector, T.; Macgregor, A.; Cassidy, A. Intakes of anthocyanins and flavones are associated with biomarkers of insulin resistance and inflammation in women. *J. Nutr.* **2014**, *144*, 202–208. [CrossRef] [PubMed]
7. Wedick, N.M.; Pan, A.; Cassidy, A.; Rimm, E.B.; Sampson, L.; Rosner, B.; Willett, W.; Hu, F.B.; Sun, Q.; van Dam, R.M. Dietary flavonoid intakes and risk of type 2 diabetes in US men and women. *Am. J. Clin. Nutr.* **2012**, *95*, 925–933. [CrossRef] [PubMed]

8. Faria, A.; Oliveira, J.; Neves, P.; Gameiro, P.; Santos-Buelga, C.; de Freitas, V.; Mateus, N. Antioxidant properties of prepared blueberry (*Vaccinium myrtillus*) extracts. *J. Agric. Food Chem.* **2005**, *53*, 6896–6902. [CrossRef] [PubMed]
9. Hosseinian, F.S.; Beta, T. Saskatoon and wild blueberries have higher anthocyanin contents than other manitoba berries. *J. Agric. Food Chem.* **2007**, *55*, 10832–10838. [CrossRef] [PubMed]
10. Youdim, K.A.; Shukitt-Hale, B.; MacKinnon, S.; Kalt, W.; Joseph, J.A. Polyphenolics enhance red blood cell resistance to oxidative stress: In vitro and in vivo. *Biochim. Biophys. Acta* **2000**, *1523*, 117–122. [CrossRef]
11. Williams, C.A.; Grayer, R.J. Anthocyanins and other flavonoids. *Nat. Prod. Rep.* **2004**, *21*, 539–573. [CrossRef] [PubMed]
12. Wu, X.; Beecher, G.R.; Holden, J.M.; Haytowitz, D.B.; Gebhardt, S.E.; Prior, R.L. Concentrations of anthocyanins in common foods in the United States and estimation of normal consumption. *J. Agric. Food Chem.* **2006**, *54*, 4069–4075. [CrossRef] [PubMed]
13. Mizuno, C.S.; Rimando, A.M. Blueberries and metabolic syndrome. *Silpakorn Univ. Sci. Technol. J.* **2009**, *3*, 2.
14. DeFuria, J.; Bennett, G.; Strissel, K.J.; Perfield, J.W., II; Milbury, P.E.; Greenberg, A.S.; Obin, M.S. Dietary blueberry attenuates whole-body insulin resistance in high fat-fed mice by reducing adipocyte death and its inflammatory sequelae. *J. Nutr.* **2009**, *139*, 1510–1516. [CrossRef] [PubMed]
15. Elks, C.M.; Terrebonne, J.D.; Ingram, D.K.; Stephens, J.M. Blueberries improve glucose tolerance without altering body composition in obese postmenopausal mice. *Obesity* **2015**, *23*, 573–580. [CrossRef] [PubMed]
16. Nair, A.R.; Elks, C.M.; Vila, J.; Del Piero, F.; Paulsen, D.B.; Francis, J. A blueberry-enriched diet improves renal function and reduces oxidative stress in metabolic syndrome animals: Potential mechanism of TLR4-MAPK signaling pathway. *PLoS ONE* **2014**, *9*, e111976. [CrossRef] [PubMed]
17. Roopchand, D.E.; Kuhn, P.; Rojo, L.E.; Lila, M.A.; Raskin, I. Blueberry polyphenol-enriched soybean flour reduces hyperglycemia, body weight gain and serum cholesterol in mice. *Pharmacol. Res.* **2013**, *68*, 59–67. [CrossRef] [PubMed]
18. Seymour, E.M.; Tanone, I.I.; Urcuyo-Llanes, D.E.; Lewis, S.K.; Kirakosyan, A.; Kondoleon, M.G.; Kaufman, P.B.; Bolling, S.F. Blueberry intake alters skeletal muscle and adipose tissue peroxisome proliferator-activated receptor activity and reduces insulin resistance in obese rats. *J. Med. Food* **2011**, *14*, 1511–1518. [CrossRef] [PubMed]
19. Takikawa, M.; Inoue, S.; Horio, F.; Tsuda, T. Dietary anthocyanin-rich bilberry extract ameliorates hyperglycemia and insulin sensitivity via activation of AMP-activated protein kinase in diabetic mice. *J. Nutr.* **2010**, *140*, 527–533. [CrossRef] [PubMed]
20. Vuong, T.; Benhaddou-Andaloussi, A.; Brault, A.; Harbilas, D.; Martineau, L.C.; Vallerand, D.; Ramassamy, C.; Matar, C.; Haddad, P.S. Antiobesity and antidiabetic effects of biotransformed blueberry juice in KKA^y mice. *Int. J. Obes. (Lond)* **2009**, *33*, 1166–1173. [CrossRef] [PubMed]
21. Wu, T.; Tang, Q.; Gao, Z.; Yu, Z.; Song, H.; Zheng, X.; Chen, W. Blueberry and mulberry juice prevent obesity development in C57BL/6 mice. *PLoS ONE* **2013**, *8*, e77585. [CrossRef] [PubMed]
22. Hoggard, N.; Cruickshank, M.; Moar, K.M.; Bestwick, C.; Holst, J.J.; Russell, W.; Horgan, G. A single supplement of a standardised bilberry (*Vaccinium myrtillus* L.) extract (36% wet weight anthocyanins) modifies glycaemic response in individuals with type 2 diabetes controlled by diet and lifestyle. *J. Nutr. Sci.* **2013**, *2*, e22. [CrossRef] [PubMed]
23. Li, D.; Zhang, Y.; Liu, Y.; Sun, R.; Xia, M. Purified anthocyanin supplementation reduces dyslipidemia, enhances antioxidant capacity, and prevents insulin resistance in diabetic patients. *J. Nutr.* **2015**, *145*, 742–748. [CrossRef] [PubMed]
24. Rebello, C.J.; Burton, J.; Heiman, M.; Greenway, F.L. Gastrointestinal microbiome modulator improves glucose tolerance in overweight and obese subjects: A randomized controlled pilot trial. *J. Diabetes Complicat.* **2015**, *29*, 1272–1276. [CrossRef] [PubMed]
25. Stull, A.J.; Cash, K.C.; Johnson, W.D.; Champagne, C.M.; Cefalu, W.T. Bioactives in blueberries improve insulin sensitivity in obese, insulin-resistant men and women. *J. Nutr.* **2010**, *140*, 1764–1768. [CrossRef] [PubMed]
26. Martineau, L.C.; Couture, A.; Spoor, D.; Benhaddou-Andaloussi, A.; Harris, C.; Meddah, B.; Leduc, C.; Burt, A.; Vuong, T.; Le Mai, P.; et al. Anti-diabetic properties of the Canadian lowbush blueberry *Vaccinium angustifolium* ait. *Phytomedicine* **2006**, *13*, 612–623. [CrossRef] [PubMed]

27. Mykkanen, O.T.; Huotari, A.; Herzig, K.H.; Dunlop, T.W.; Mykkanen, H.; Kirjavainen, P.V. Wild blueberries (*Vaccinium myrtillus*) alleviate inflammation and hypertension associated with developing obesity in mice fed with a high-fat diet. *PLoS ONE* **2014**, *9*, e114790. [CrossRef] [PubMed]
28. Prior, R.L.; Wu, X.; Gu, L.; Hager, T.J.; Hager, A.; Howard, L.R. Whole berries versus berry anthocyanins: Interactions with dietary fat levels in the C57BL/6J mouse model of obesity. *J. Agric. Food Chem.* **2008**, *56*, 647–653. [CrossRef] [PubMed]
29. Vendrame, S.; Zhao, A.; Merrow, T.; Klimis-Zacas, D. The effects of wild blueberry consumption on plasma markers and gene expression related to glucose metabolism in the obese Zucker rat. *J. Med. Food* **2015**, *18*, 619–624. [CrossRef] [PubMed]
30. Basu, A.; Du, M.; Leyva, M.J.; Sanchez, K.; Betts, N.M.; Wu, M.; Aston, C.E.; Lyons, T.J. Blueberries decrease cardiovascular risk factors in obese men and women with metabolic syndrome. *J. Nutr.* **2010**, *140*, 1582–1587. [CrossRef] [PubMed]
31. Stull, A.J.; Cash, K.C.; Champagne, C.M.; Gupta, A.K.; Boston, R.; Beyl, R.A.; Johnson, W.D.; Cefalu, W.T. Blueberries improve endothelial function, but not blood pressure, in adults with metabolic syndrome: A randomized, double-blind, placebo-controlled clinical trial. *Nutrients* **2015**, *7*, 4107–4123. [CrossRef] [PubMed]
32. Riccardi, G.; Rivellese, A.A. Effects of dietary fiber and carbohydrate on glucose and lipoprotein metabolism in diabetic patients. *Diabetes Care* **1991**, *14*, 1115–1125. [CrossRef] [PubMed]
33. Cebeci, F.; Sahin-Yesilcubuk, N. The matrix effect of blueberry, oat meal and milk on polyphenols, antioxidant activity and potential bioavailability. *Int. J. Food Sci. Nutr.* **2014**, *65*, 69–78. [CrossRef] [PubMed]
34. Leenen, R.; Roodenburg, A.J.C.; Tijburg, L.B.M.; Wiseman, S.A. A single dose of tea with or without milk increases plasma antioxidant activity in humans. *Eur. J. Clin. Nutr.* **2000**, *54*, 87–92. [CrossRef] [PubMed]
35. Reddy, V.C.; Sagar, G.V.V.; Sreeramulu, D.; Venu, L.; Raghunath, M. Addition of milk does not alter the antioxidant activity of black tea. *Ann. Nutr. MeTable* **2005**, *49*, 189–195. [CrossRef] [PubMed]
36. Serafini, M.; Ghiselli, A.; Ferro-Luzzi, A. In vivo antioxidant effect of green and black tea in man. *Eur. J. Clin. Nutr.* **1996**, *50*, 28–32. [PubMed]
37. Van het Hof, K.H.; Kivits, G.A.A.; Weststrate, J.A.; Tijburg, L.B.M. Bioavailability of catechins from tea: The effect of milk. *Eur. J. Clin. Nutr.* **1998**, *52*, 356–359. [CrossRef] [PubMed]
38. Shoelson, S.E.; Lee, J.; Goldfine, A.B. Inflammation and insulin resistance. *J. Clin. Investig.* **2006**, *116*, 1793–1801. [CrossRef] [PubMed]
39. Prior, R.L.; Wu, X.; Gu, L.; Hager, T.; Hager, A.; Wilkes, S.; Howard, L. Purified berry anthocyanins but not whole berries normalize lipid parameters in mice fed an obesogenic high fat diet. *Mol. Nutr. Food Res.* **2009**, *53*, 1406–1418. [CrossRef] [PubMed]
40. Weisberg, S.P.; McCann, D.; Desai, M.; Rosenbaum, M.; Leibel, R.L.; Ferrante, A.W., Jr. Obesity is associated with macrophage accumulation in adipose tissue. *J. Clin. Investig.* **2003**, *112*, 1796–1808. [CrossRef] [PubMed]
41. Xu, H.; Barnes, G.T.; Yang, Q.; Tan, G.; Yang, D.; Chou, C.J.; Sole, J.; Nichols, A.; Ross, J.S.; Tartaglia, L.A.; et al. Chronic inflammation in fat plays a crucial role in the development of obesity-related insulin resistance. *J. Clin. Investig.* **2003**, *112*, 1821–1830. [CrossRef] [PubMed]
42. Vendrame, S.; Daugherty, A.; Kristo, A.S.; Riso, P.; Klimis-Zacas, D. Wild blueberry (*Vaccinium angustifolium*) consumption improves inflammatory status in the obese Zucker rat model of the metabolic syndrome. *J. Nutr. Biochem.* **2013**, *24*, 1508–1512. [CrossRef] [PubMed]
43. Zhu, Y.; Ling, W.; Guo, H.; Song, F.; Ye, Q.; Zou, T.; Li, D.; Zhang, Y.; Li, G.; Xiao, Y.; et al. Anti-inflammatory effect of purified dietary anthocyanin in adults with hypercholesterolemia: A randomized controlled trial. *Nutr. Metab. Cardiovasc. Dis.* **2013**, *23*, 843–849. [CrossRef] [PubMed]
44. Johnson, S.A.; Figueroa, A.; Navaei, N.; Wong, A.; Kalfon, R.; Ormsbee, L.T.; Feresin, R.G.; Elam, M.L.; Hooshmand, S.; Payton, M.E.; et al. Daily blueberry consumption improves blood pressure and arterial stiffness in postmenopausal women with pre- and stage 1-hypertension: A randomized, double-blind, placebo-controlled clinical trial. *J. Acad. Nutr. Diet.* **2015**, *115*, 369–377. [CrossRef] [PubMed]
45. Riso, P.; Klimis-Zacas, D.; Del Bo, C.; Martini, D.; Campolo, J.; Vendrame, S.; Moller, P.; Loft, S.; de Maria, R.; Porrini, M. Effect of a wild blueberry (*Vaccinium angustifolium*) drink intervention on markers of oxidative stress, inflammation and endothelial function in humans with cardiovascular risk factors. *Eur. J. Nutr.* **2013**, *52*, 949–961. [CrossRef] [PubMed]

46. Vuong, T.; Martineau, L.C.; Ramassamy, C.; Matar, C.; Haddad, P.S. Fermented canadian lowbush blueberry juice stimulates glucose uptake and amp-activated protein kinase in insulin-sensitive cultured muscle cells and adipocytes. *Can. J. Physiol. Pharmacol.* **2007**, *85*, 956–965. [CrossRef] [PubMed]

47. Kato, M.; Tani, T.; Terahara, N.; Tsuda, T. The anthocyanin delphinidin 3-rutinoside stimulates glucagon-like peptide-1 secretion in murine GLU Tag cell line via the Ca^{2+}/calmodulin-dependent kinase II pathway. *PLoS ONE* **2015**, *10*, e0126157. [CrossRef] [PubMed]

48. Clemmensen, C.; Smajilovic, S.; Smith, E.P.; Woods, S.C.; Brauner-Osborne, H.; Seeley, R.J.; D'Alessio, D.A.; Ryan, K.K. Oral l-arginine stimulates GLP-1 secretion to improve glucose tolerance in male mice. *Endocrinology* **2013**, *154*, 3978–3983. [CrossRef] [PubMed]

49. Rocca, A.S.; LaGreca, J.; Kalitsky, J.; Brubaker, P.L. Monounsaturated fatty acid diets improve glycemic tolerance through increased secretion of glucagon-like peptide-1. *Endocrinology* **2001**, *142*, 1148–1155. [CrossRef] [PubMed]

50. Tolhurst, G.; Zheng, Y.; Parker, H.E.; Habib, A.M.; Reimann, F.; Gribble, F.M. Glutamine triggers and potentiates glucagon-like peptide-1 secretion by raising cytosolic Ca^{2+} and camp. *Endocrinology* **2011**, *152*, 405–413. [CrossRef] [PubMed]

51. Scazzocchio, B.; Vari, R.; Filesi, C.; D'Archivio, M.; Santangelo, C.; Giovannini, C.; Iacovelli, A.; Silecchia, G.; Li Volti, G.; Galvano, F.; et al. Cyanidin-3-*O*-β-glucoside and protocatechuic acid exert insulin-like effects by upregulating ppargamma activity in human omental adipocytes. *Diabetes* **2011**, *60*, 2234–2244. [CrossRef] [PubMed]

52. Vendrame, S.; Daugherty, A.; Kristo, A.S.; Klimis-Zacas, D. Wild blueberry (*Vaccinium angustifolium*)-enriched diet improves dyslipidaemia and modulates the expression of genes related to lipid metabolism in obese Zucker rats. *Br. J. Nutr.* **2014**, *111*, 194–200. [CrossRef] [PubMed]

53. Xia, M.; Hou, M.; Zhu, H.; Ma, J.; Tang, Z.; Wang, Q.; Li, Y.; Chi, D.; Yu, X.; Zhao, T.; et al. Anthocyanins induce cholesterol efflux from mouse peritoneal macrophages: The role of the peroxisome proliferator-activated receptor γ-liver X receptor α-ABCA1 pathway. *J. Biol. Chem.* **2005**, *280*, 36792–36801. [CrossRef] [PubMed]

54. Nehlin, J.O.; Mogensen, J.P.; Petterson, I.; Jeppesen, L.; Fleckner, J.; Wulff, E.M.; Sauerberg, P. Selective PPAR agonists for the treatment of type 2 diabetes. *Ann. N. Y. Acad. Sci.* **2006**, *1067*, 448–453. [CrossRef] [PubMed]

55. Liu, J.; Gao, F.; Ji, B.; Wang, R.; Yang, J.; Liu, H.; Zhou, F. Anthocyanins-rich extract of wild Chinese blueberry protects glucolipotoxicity-induced INS832/13 β-cell against dysfunction and death. *J. Food Sci. Technol.* **2015**, *52*, 3022–3029. [CrossRef] [PubMed]

56. Poitout, V.; Robertson, R.P. Glucolipotoxicity: Fuel excess and β-cell dysfunction. *Endocr. Rev.* **2008**, *29*, 351–366. [CrossRef] [PubMed]

57. Zhu, W.; Jia, Q.; Wang, Y.; Zhang, Y.; Xia, M. The anthocyanin cyanidin-3-*O*-β-glucoside, a flavonoid, increases hepatic glutathione synthesis and protects hepatocytes against reactive oxygen species during hyperglycemia: Involvement of a cAMP-PKA-dependent signaling pathway. *Free Radic. Biol. Med.* **2012**, *52*, 314–327. [CrossRef] [PubMed]

58. Roy, M.; Sen, S.; Chakraborti, A.S. Action of pelargonidin on hyperglycemia and oxidative damage in diabetic rats: Implication for glycation-induced hemoglobin modification. *Life Sci.* **2008**, *82*, 1102–1110. [CrossRef] [PubMed]

antioxidants

MDPI

Article

Blueberry Consumption Affects Serum Uric Acid Concentrations in Older Adults in a Sex-Specific Manner

Carol L. Cheatham [1,2,*], Itzel Vazquez-Vidal [2], Amanda Medlin [3] and V. Saroja Voruganti [2,4]

[1] Department of Psychology & Neuroscience, University of North Carolina at Chapel Hill, Chapel Hill, NC 27599, USA
[2] Nutrition Research Institute, University of North Carolina at Chapel Hill, 500 Laureate Way, Rm 1101, Kannapolis, NC 28081, USA; itzel_vazquez@unc.edu (I.V.-V.); saroja@unc.edu (V.S.V.)
[3] University of North Carolina at Charlotte, Charlotte, NC 28223, USA; amedli13@uncc.edu
[4] Department of Nutrition, University of North Carolina at Chapel Hill, Chapel Hill, NC 27599, USA
* Correspondence: carol_cheatham@unc.edu; Tel.: +1-704-250-5010; Fax: +1-704-250-5001

Academic Editor: Dorothy Klimis-Zacas
Received: 1 August 2016; Accepted: 18 November 2016; Published: 29 November 2016

Abstract: Blueberries are rich in antioxidants and may protect against disease. Uric acid accounts for about 50% of the antioxidant properties in humans. Elevated levels of serum uric acid (SUA) or hyperuricemia is a risk factor for cardiovascular disease (CVD). The aim was to determine the effect of blueberries on SUA in older adults. Participants ($n = 133$, 65–80 years) experiencing mild cognitive impairment (MCI) were randomized in a double-blind 6-month clinical trial to either blueberry or placebo. A reference group with no MCI received no treatment. The mean (SD) SUA at baseline were 5.45 (0.9), 6.4 (1.3) and 5.8 (1.4) mg/dL in reference, placebo, and treatment groups, respectively. Baseline SUA was different in men and women (6.25 (1.1) vs. 5.35 (1.1), $p = 0.001$). During the first three months, SUA decreased in the blueberry group and was significantly different from the placebo group in both men and women ($p < 0.0003$). Sex-specific differences became apparent after 3 months, when only men showed an increase in SUA in the blueberry group and not in the placebo ($p = 0.0006$) between 3 and 6 months. At 6 months SUA had rebounded in both men and women and returned to baseline levels. Baseline SUA was correlated with CVD risk factors, waist circumference and triglycerides ($p < 0.05$), but differed by sex. Overall, 6 m SUA changes were negatively associated with triglycerides in men, but not in women. Group-wise association between 6 m SUA changes and CVD risk factors showed associations with diastolic blood pressure, triglycerides and high-density lipoprotein (HDL) cholesterol in women of the Blueberry group but not in men or any sex in the placebo group. In summary, blueberries may affect SUA and its relationship with CVD risk in a sex-specific manner.

Keywords: hyperuricemia; blueberries; cardiovascular disease; antioxidant

1. Introduction

Uric acid accounts for more than half of the antioxidative properties in humans [1–4]. However, elevated serum uric acid concentrations (SUA) or hyperuricemia is considered a risk factor for the development of oxidative stress and/or inflammation-related diseases such as gout, hypertension, cardiovascular disease (CVD) and chronic kidney disease (CKD) [1]. Uric acid has a paradoxical role in human metabolism: on one hand, it acts as an antioxidant in hydrophilic environment, whereas on the other hand, it can propagate a chain reaction and cause oxidative damage to cells in reaction with other radicals [1]. However, the role of SUA in CVD remains controversial with conflicting information being put forward regarding the link between SUA and CVD [5–7].

Plant polyphenols have been shown to be protective against diseases such as certain cancers, CVD, diabetes, and aging [8–10]. Blueberries contain anthocyanins, which are recognized to possess antioxidant and anti-inflammatory properties. Human and animal studies have shown beneficial effects of blueberry consumption on lifestyle-related chronic diseases [11]. Studies have been conducted to understand the effect of polyphenol and anthocyanins containing foods on circulating uric acid concentrations.On the one hand, red wine has been shown to acutely increase plasma urate concentrations [12], whereas, on the other hand, dealcoholized red wine has been shown to decrease plasma uric acid concentrations [13]. Similarly, cherry consumption has been shown to reduce SUA concentrations whereas acaiberry had no effects on them [14,15]. Although, to the best of our knowledge, blueberries have not been tested for their effects on SUA and no scientific evidence links them so far, they are thought to be functional foods for curing gout, an inflammatory arthritic disease caused by hyperuricemia, mainly for their anthocyanin content or antioxidant property.

The role of berry fruits in improving cardiovascular health is gaining importance [8]. Studies have shown beneficial effects of flavonoids and anthocyanins on endothelial function, type 2 diabetes, and hypertension. A study in 93,600 young to middle-aged women from the Nurses Health Study (NHS) II showed that high intake of anthocyanins can help reduce risk for myocardial infarction [16]. Another study from the same group also showed that anthocyanins and specific flavones can prevent hypertension [17] and decrease the risk for type 2 diabetes [18]. In a randomized, double-blind placebo-controlled clinical trial, blueberry consumption was associated with improvement in endothelial function, but not in blood pressure, in 44 adult men [19]. In contrast, in another double-blind placebo-controlled clinical trial, blueberries were found to decrease blood pressure and arterial stiffness in 8 weeks in postmenopausal women [20]. Additionally, a dose and time-dependent improvement was observed in vascular function following blueberry consumption in men participating in a double-blind, crossover intervention study [21]. However, very few studies have investigated the effect of blueberry consumption on SUA concentration or its relationship with CVD risk factors. A literature review revealed only one study which was conducted in 15 healthy adults that did not find any significant differences in SUA in individuals with high, low, or no dose blueberries [22].

Thus, the aim of this study was to determine the impact of blueberries on SUA concentrations and their relation to CVD risk factors in older adults. Our central hypothesis is that the blueberries, when added to the diet, will enhance the body's antioxidant status, lower SUA concentrations, and improve related CVD risk factors.

2. Research Design and Methods

2.1. Study Population

In the context of a larger study that was designed to assess the effects of blueberries on cognitive abilities, adults aged 65 to 80 years, who were beginning to experience mild cognitive decline, but were generally healthy, were enrolled in a 6-month double-blind randomized clinical trial across a 3-year rolling enrollment. Inclusion criteria included those who consumed fewer than five daily servings of fruits and vegetables; were not diagnosed with dementia or Alzheimer's disease, central nervous system disorders, psychiatric disorders, gastrointestinal issues, digestive issues, or diabetes; had a body mass index (BMI) less than 34.9 lbs./in^2; were not taking certain medicines (e.g., medications with known cognitive side-effects); and were right-handed. The study was approved by the Institutional Review Board for the University of North Carolina at Chapel Hill, and all participants gave written and verbal informed consent (11-2075).

Intervention participants were randomly assigned to either the blueberry or placebo group (Figure 1). Participants were age-matched in half-decade categories (65–69, 70–74, and 75–79 years old). Single-year-crop wild blueberries donated by the Wild Blueberry Association of North America (Old Town, ME, USA) were freeze-dried and pulverized into a powder at Futureceuticals, Inc. (Momence, IL, USA) and packaged in 17.5 g packets at WePackItAll (Irwindale, CA, USA). The placebo powder was created by Futureceuticals from a recipe developed by the High Bush Blueberry Council and was also packaged at WePackItAll. The packaging was blank with the exception of one of four letters (C, D, E, or F), which constituted the blind. Powders were analyzed for nutrient content and antioxidant activity by Medallion Labs (Minneapolis, MN, USA) at least semi-annually during the study to insure the powders maintained consistency over time (see Table 1 for nutrient composition).

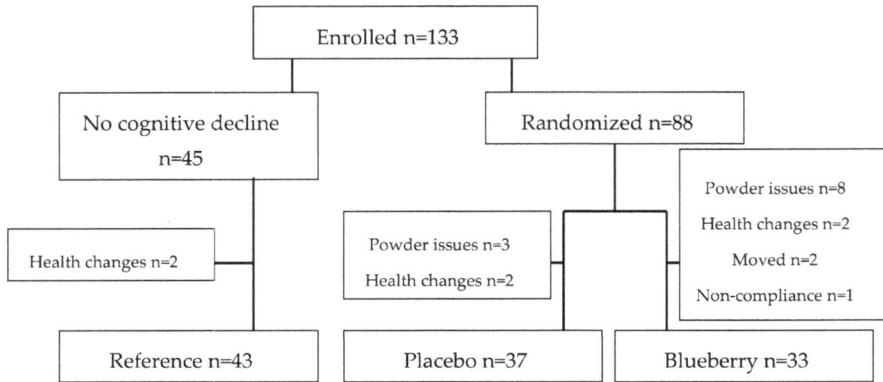

Figure 1. A Consort diagram of the participant flow of the parent clinical trial.

Table 1. Average nutrition provided by powders per day if participants were 100% compliant.

Composition	Blueberry	Placebo
Calories	139	137
Saturated Fat (g)	0.084	0.000
Monounsaturated Fatty Acids (g)	0.138	0.000
Polyunsaturated Fatty Acids (g)	0.403	0.004
Trans Fatty Acids (g)	0.003	0.000
Cholesterol (g)	0.37	0.35
Sodium (mg)	2.60	7.46
Carbohydrates (g)	32.08	34.26
Fiber (g)	7.05	0.02
Fructose (g)	11.47	4.56
Glucose (g)	10.65	0.30
Sucrose (g)	0.05	0.04
Maltose (g)	0.07	0.54
Lactose (g)	0.04	0.04
Protein (g)	1.08	0.27
Calcium (g)	37.25	3.54
Iron (g)	0.54	0.05
Vitamin C (g)	9.34	0.25

Participants were instructed to consume two packets of powder (35 g)—which for the blueberries was the equivalent of two cups of fresh blueberries—daily for 180 days. Participants were further instructed that the powders were not to be heated, cooked, or added to already hot foods; it was also recommended that they not mix the powder with dairy products [23,24]. Participants recorded powder consumption times and details in a provided food diary. Any packets not consumed were returned to researchers; compliance was calculated from returned packets.

Participants were interviewed about the foods listed in their diary using a 3-pass diet recall system. Data were collected in a standardized format with the aid of food models for accurate portion determination utilizing the Nutrition Data System for Research (NDSR) software (Minneapolis, MN, USA). The NDSR software calculates average daily caloric and nutrient intake.

2.2. Uric Acid

Twelve-hour fasted blood was obtained from participants at baseline, 3 months, and 6 months. Blood was collected by a trained phlebotomist through blood draw using ethylenediaminetetraacetic acid (EDTA) coated tubes and serum tubes. Blood was centrifuged for 15 min at 1500g at 4 °C; layers were aliquoted into separate tubes and immediately stored at −80 °C. The blood was processed by Carolinas Medical Center LabCorp (Charlotte, NC, USA); the blood was assayed for creatinine, glucose, cholesterol, triglycerides, high-density lipoprotein (HDL) cholesterol and low-density lipoprotein (LDL) cholesterol.

2.3. Physical Examination

Anthropometrics and blood pressure were collected from participants at each session by trained research assistants. After participants sat quietly for five min with both feet on the floor, blood pressure was measured using an Omron HEM-907XL digital blood pressure monitor. Waist circumference was measured 1 inch above the navel using a soft tape measure. Participants' weights and heights were measured without shoes and after emptying pockets and removing excess clothing. The Cardinal Detecto Pro Doc Series digital physician scale was used to measure weight. Height was measured to the nearest quarter inch using a Charder portable stadiometer situated against the wall.

2.4. Statistical Analysis

All statistical analyses were computed using STATA, version 14.0 (StataCorp LP, College Station, TX, USA). A test of normality was performed and the variables showing skewed distribution were log transformed to meet the normal distribution. t-tests and ANOVA were used to analyze group and sex differences. Pearson correlation tests and multiple linear regression analyses were used to assess the relationship between SUA concentrations and CVD risk factors. A value of <0.05 (two-tailed) was considered to be statistically significant after adjustment for multiple tests.

3. Results

A total of 58 men and 75 women aged 65 to 80 years participated in the larger study. Of the 133 enrolled participants, 107 provided blood samples and were included in the analysis: 47 men and 60 women. Their descriptive characteristics are depicted in Table 2. The 6-month period treatment showed no differences in body weight and waist circumferences in any of the three groups. No other metabolic risk factors showed any significant changes across the 6-month period.

Table 2. Descriptive statistics at baseline, 3 months, and 6 months.

Trait	Baseline	Mean (SD) 3 Months	6 Months
Age	72.68 ± 4.3		
Body weight (lb)	170.17 ± 31.7	171.43 ± 31.2	171.52 ± 31.5
Waist circumference (in)	39.60 ± 4.0	39.86 ± 4.2	39.61 ± 4.0
Systolic blood pressure (mmHg)	131.76 ± 17.3	129.47 ± 16.3	130.59 ± 16.9
Diastolic blood pressure (mmHg)	74.42 ± 9.1	73.68 ± 9.2	73.44 ± 10.9
Glucose (mg/dL)	100.22 ± 11.5	98.97 ± 13.3	99.88 ± 12.6
Triglycerides (mg/dL)	137.29 ± 97.7	131.28 ± 69.2	127.82 ± 64.4
Total cholesterol (mg/dL)	177.98 ± 34.4	179.48 ± 33.0	177.70 ± 36.7
LDL cholesterol (mg/dL)	98.50 ± 28.3	99.00 ± 27.0	97.10 ± 30.6
HDL cholesterol (mg/dL)	53.64 ± 16.0	54.73 ± 16.5	54.68 ± 16.5
SUA (mg/dL)	5.75 ± 1.3	5.21 ± 1.4	5.66 ± 1.3

LDL-low-density lipoprotein: HDL-high-density lipoprotein: SD-standard deviation; SUA-serum uric acid.

3.1. Compliance

In both treatment groups, although not significant, women seem to be more compliant with powder consumption than men. In the placebo group, compliance was 93% and 81% in men at three and six months, respectively, whereas it was 97% in women at both the three- and six-month timepoints. In the blueberry group, however, compliance was lower at 78% and 70% in men at three and six months, respectively and 81% and 62% in women at three and six months, respectively.

3.2. Changes in SUA Following Supplementation with Blueberries or Placebo

At baseline mean (SD) SUA, levels were 5.91 (1.3) mg/dL. When categorized by sex and treatment groups, SUA concentrations were higher in men than women (6.25 (1.1) vs. 5.35 (1.1) mg/dL, $p = 0.001$) but not significantly different between placebo and treatment groups (6.1 (1.2) vs. 5.8 (1.4), $p = 0.43$) at baseline. Changes in SUA between baseline and 3 months (0–3 months), 3 to 6 months (3–6 months) and baseline to 6 months (0–6 months) were adjusted for age, waist circumference, total calorie intake and blueberry intake compliance and residuals were used for further analysis. Figure 2 depicts the differences in these changes between the two sexes and groups. SUA during 0–3 months decreased significantly in the blueberry group as compared to the placebo group in both men ($p = 0.0002$) and women ($p = 0.00001$). However, during 3–6 months, in men, SUA increased significantly in the blueberry group as compared to the placebo group ($p = 0.0006$) but not in women ($p = 0.87$). By the end of 6 months, SUA had decreased to its baseline levels and was not significantly different between groups in both men ($p = 0.23$) and women ($p = 0.54$).

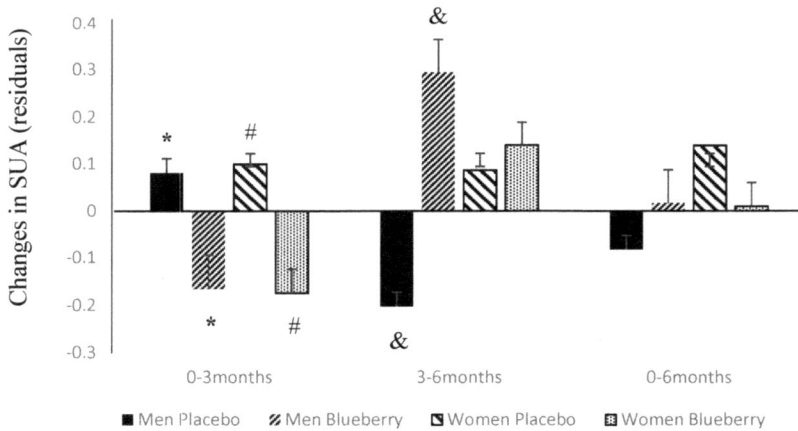

Figure 2. SUA (serum uric acid) changes over 6 months according to sex and treatment group. Residual values are shown as mean (SD); * $p = 0.0002$, # $p = 0.00001$, & $p = 0.0006$.

3.3. Changes in SUA and Its Relation to Changes in Other CVD Risk Factors

Regardless of sex differences, baseline SUA concentration was significantly correlated with waist circumference ($r = 0.36$, $p < 0.001$), triglycerides ($r = 0.31$, $p < 0.01$), glucose ($r = 0.27$, $p < 0.01$), and HDL cholesterol ($r = -0.43$, $p < 0.001$). Sex-specific significant correlations of SUA concentrations were observed with triglycerides ($r = 0.30$, $p < 0.05$) in men and with waist circumference ($r = 0.50$, $p < 0.0001$), triglycerides ($r = 0.42$, $p < 0.001$), glucose ($r = 0.45$, $p = < 0.001$), and HDL cholesterol ($r = -0.42$, $p = 0.001$) in women. Changes in SUA concentrations (baseline to 6 months) were positively correlated with glucose ($r = 0.32$, $p = 0.03$) and negatively with systolic ($r = -0.36$, $p = 0.009$) and diastolic blood pressure ($r = -0.36$, $p = 0.009$) in women (data not shown in tables). Multiple regression, adjusted for blueberry compliance, showed the baseline SUA concentrations were significantly correlated with CVD risk factors, waist circumference and serum triglycerides ($p < 0.05$). However, the association of changes in SUA concentrations differed by sex and treatment. Changes in SUA concentrations were negatively associated with blood glucose and triglycerides in men whereas no significant associations were observed in women (Table 3).

Table 3. Multiple regression analysis of baseline SUA concentrations and changes in SUA concentrations with cardiovascular disease risk factors.

Trait **	CVD Risk Factor	β (SE)	t	p Value *
Baseline—All	Waist circumference	0.073 (0.03)	2.67	**0.009**
	Glucose	0.009 (0.009)	0.96	0.34
	Systolic blood pressure	−0.003 (0.008)	−0.42	0.67
	Diastolic blood pressure	−0.0004 (0.01)	−0.03	0.98
	Triglycerides	0.003 (0.001)	2.31	**0.023**
	Total cholesterol	−0.003 (0.01)	0.44	0.66
	LDL cholesterol	0.005 (0.01)	−0.30	0.77
	HDL cholesterol	−0.010 (0.10)	0.39	0.24
Men	Waist circumference	0.0102 (0.06)	0.18	0.86
	Glucose	−0.002 (0.02)	−0.15	0.88
	Systolic blood pressure	−0.16 (0.01)	−1.27	0.21
	Diastolic blood pressure	0.24 (0.02)	1.08	0.29
	Triglycerides	0.002 (0.002)	1.51	0.14
	Total cholesterol	−0.002 (0.02)	−0.13	0.90
	LDL cholesterol	−0.002 (0.01)	0.12	0.91
	HDL cholesterol	−0.002 (0.02)	−0.15	0.88
Women	Waist circumference	0.085 (0.03)	2.72	**0.009**
	Glucose	0.018 (0.01)	1.33	0.19
	Systolic blood pressure	0.003 (0.01)	0.25	0.80
	Diastolic blood pressure	−0.173 (0.02)	−0.95	0.35
	Triglycerides	0.002 (0.01)	−0.78	0.44
	Total cholesterol	−0.004 (0.01)	−0.31	0.76
	LDL cholesterol	−0.008 (0.02)	0.51	0.61
	HDL cholesterol	−0.09 (0.1)	1.33	0.19
Changes over 6-month period **—All	Waist circumference	0.004 (0.1)	0.08	0.94
	Glucose	0.014 (0.01)	0.96	0.34
	Systolic blood pressure	−0.009 (0.01)	−0.84	0.41
	Diastolic blood pressure	−0.0003 (0.01)	−0.02	0.98
	Triglycerides	−0.0003 (0.002)	−0.17	0.87
	Total cholesterol	−0.008 (0.02)	−0.50	0.62
	LDL cholesterol	0.021 (0.02)	1.1	0.28
	HDL cholesterol	0.036 (0.02)	1.58	0.12
Men	Waist circumference	0.083 (0.05)	1.6	0.14
	Glucose	−0.045 (0.02)	−2.19	**0.05**
	Systolic blood pressure	−0.022 (0.02)	−1.33	0.21
	Diastolic blood pressure	0.022 (0.02)	1.23	0.24
	Triglycerides	−0.007 (0.002)	−3.21	**0.008**
	Total cholesterol	−0.0001 (0.02)	−0.01	0.99
	LDL cholesterol	0.009 (0.03)	0.34	0.74
	HDL cholesterol	0.070 (0.04)	1.81	**0.098**
Women	Waist circumference	−0.062 (0.07)	−0.87	0.40
	Glucose	0.033 (0.02)	2.02	**0.06**
	Systolic blood pressure	−0.005 (0.1)	−0.48	0.64
	Diastolic blood pressure	−0.033 (0.03)	−1.32	0.21
	Triglycerides	0.002 (0.005)	0.33	0.74
	Total cholesterol	−0.025 (0.02)	−1.32	0.20
	LDL cholesterol	0.042 (0.02)	−1.78	**0.08**
	HDL cholesterol	0.025 (0.02)	0.63	0.54

* all p values ≤ 0.06 are bolded; ** adjusted for age, compliance and total calorie intake. CVD: cardiovascular disease. SE: Standard error of mean.

When categorized by treatment groups (blueberry vs. placebo), changes in SUA concentrations were not associated with any CVD risk factors in the placebo group. Conversely, in the blueberry group, changes in SUA concentrations were significantly associated with diastolic blood pressure, and serum levels of triglycerides and HDL cholesterol and showed suggestive evidence of association with changes in total cholesterol in women but not in men (Table 4). At baseline, SUA, total cholesterol and changes in SUA and total cholesterol were similar in both sexes in the placebo group. However, in the blueberry group, beta coefficient was in opposite direction for association between changes in SUA and diastolic blood pressure, triglycerides, HDL and total cholesterol between men and women. Thus, there seems to be an effect modification by sex in the relationship of SUA with these CVD risk factors.

Table 4. Multiple regression analysis of changes in SUA concentrations with cardiovascular disease risk factors in the treatment groups.

Trait **	CVD Risk Factor	β (SE)	*t*	*p* Value *
Placebo group—Men	Waist circumference	−0.036 (0.06)	−0.63	0.59
	Glucose	0.029 (0.14)	0.21	0.86
	Systolic blood pressure	−0.00009 (0.01)	0.02	0.99
	Diastolic blood pressure	0.006 (0.02)	0.29	0.80
	Triglycerides	0.0004 (0.001)	0.36	0.75
	Total cholesterol	−0.003 (0.008)	−0.40	0.73
	LDL cholesterol	−0.004 (0.007)	−0.59	0.57
	HDL cholesterol	−0.047 (0.05)	−0.86	0.48
Women	Waist circumference	−0.107 (0.08)	−1.37	0.21
	Glucose	0.0034 (0.01)	0.19	0.85
	Systolic blood pressure	−0.0105 (0.01)	−0.74	0.48
	Diastolic blood pressure	0.0031 (0.02)	0.14	0.90
	Triglycerides	0.0047 (0.007)	0.72	0.50
	Total cholesterol	−0.0163 (0.02)	−0.96	0.37
	LDL cholesterol	0.002 (0.009)	0.25	0.81
	HDL cholesterol	0.0043 (0.04)	0.11	0.91
Blueberry group—Men	Waist circumference	0.007 (0.04)	0.18	0.86
	Glucose	0.002 (0.06)	0.03	0.98
	Systolic blood pressure	−0.022 (0.01)	−0.19	0.86
	Diastolic blood pressure	−0.0003 (0.01)	−0.02	0.99
	Triglycerides	0.0025 (0.002)	1.21	0.28
	Total cholesterol	−0.004 (0.007)	−0.56	0.60
	LDL cholesterol	−0.003 (0.005)	−0.05	0.96
	HDL cholesterol	0.077 (0.04)	2.12	0.09
Women	Waist circumference	−0.029 (0.10)	−1.91	0.15
	Glucose	−0.198 (0.01)	−2.05	0.13
	Systolic blood pressure	0.0052 (0.003)	1.65	0.20
	Diastolic blood pressure	0.045 (0.008)	5.38	**0.01**
	Triglycerides	−0.015 (0.003)	−4.87	**0.02**
	Total cholesterol	0.011 (0.004)	2.98	**0.06**
	LDL cholesterol	−0.001 (0.003)	−0.42	0.68
	HDL cholesterol	−0.117 (0.02)	−5.83	**0.01**

* *p* values ≤ 0.06 are bolded; ** adjusted for age, total calorie intake and compliance.

4. Discussion

In the present study, we investigated the effects of blueberry consumption on SUA levels and its relationship with CVD risk factors in older adults. Uric acid, a major antioxidant in humans, can also act as pro-oxidant and increase the risk for major chronic diseases such as hypertension, type 2 diabetes, CVD, and CKD thus has contradictory effects on metabolism. Elevated SUA concentrations have been associated with increased risk for all-cause and CVD mortality [25]. A number of studies have investigated the role of uric acid in the development of CVD, however, very limited studies have been conducted in older adults. Moreover, no studies have investigated the effect of nutritional intervention (blueberry) on SUA concentrations in older individuals. Considering that both have anti-oxidative properties, it is important that we understand the effect of blueberry supplementation on SUA concentrations.

Our results show that, at baseline, men had significantly higher levels of SUA than women. Significant differences have been observed in the levels of SUA concentrations in men and women. Several studies have reported higher levels of SUA in men as compared to women [26–33], which is similar to our results. Similarly, sex-specific differences have also been reported with respect to association of SUA with CVD risk factors. Chou et al. [34] conducted a stratified analysis and found association of SUA with insulin resistance and plasma glucose levels in women but not in men. Previous studies from our group and others have shown a positive association between SUA and waist circumference [27,28,32,34]. In the current study, we found a strong correlation between SUA and waist circumference at baseline, mainly in women.

Many studies have observed the association between CVD risk factors and SUA concentrations to be stronger in women than men [32,33,35]. We also observed strong correlations of baseline SUA with serum triglycerides. This, again, is a replication of observations reported by several studies [32,35–37]. A retrospective analysis of the database from the Laboratory Information System—a database of data from a cohort of outpatient adults referred by general practitioners for routine medical check-ups for three years from 2005–2008—showed that triglycerides were significantly associated with circulating uric acid in women but not in men [35]. Similarly, a study conducted in elderly adults reported that high SUA levels predicted metabolic syndrome in older women but not men [31]. In our study, we observed a similar pattern of associations. Changes in SUA between baseline and the 6-month time-period were significantly associated with CVD risk factors (glucose and triglycerides) in men but not women. Changes in SUA regardless of treatment group also showed sex-specific differences in associations between SUA and CVD risk factors. Thus, our study replicates and confirms that the association of baseline SUA and changes in SUA are significantly associated with some CVD risk factors in a sex-specific manner.

As a next step, we investigated whether blueberry consumption affected SUA concentrations. We found that SUA concentrations initially decreased regardless of sex differences. However, after the 3-month time point, sex-specific differences became apparent. Significant differences between placebo and blueberry groups between 3 and 6 months were found only in men and not in women. Likewise, changes in SUA concentration were significantly associated with diastolic blood pressure, and serum levels of triglycerides and HDL cholesterol only in women but not men in the blueberry group. No significant associations or sex-specific differences were observed in the placebo group. However, the sex-specific differences in changes in SUA between and within groups are noteworthy. To the best of our knowledge, this is the first study to focus on the effects of blueberry on SUA concentrations over a 6-month long period. There is only one other study [22] in which scientists investigated the effects of high and low dose of blueberries on measures of antioxidant status and included SUA as one of the measures. However, this study measured only the acute effects (fasting, 1, 2, and 3 h after sample consumption) and did not find any significant differences in SUA concentrations. We, on the other hand, investigated the long-term effects of blueberry consumption. We also found association of blueberry compliance with SUA concentrations where increasing compliance was associated with deceasing SUA in women.

Our study has some limitations. Most importantly, our sample size may be too small to detect a significant effect: the study was not powered for a sex split. Despite the small size, we were able to find a significant association between changes in SUA, consumption of blueberries, and CVD risk factors. Now that the evidence for sex differences is clear, future studies should include sufficient numbers of participants to analyze by sex. In addition, eleven participants dropped out because of issues with the study powders. Thus, the results are only generalizable to those who can tolerate a diet with added berries and sugars. Finally, we didn't have measures of individual markers of anti-oxidant status, and as a result, we could not evaluate the direct effect of blueberry consumption on anti-oxidant status.

5. Conclusions

In conclusion, blueberry consumption seems to affect SUA concentrations and its relationship with CVD risk factors in a sex-specific manner. Importantly, lowering of SUA with consumption of blueberry powder in women, and not men, warrants further studies to confirm and validate these results in a larger sample.

Acknowledgments: We would like to thank the families that gave their time and energy so selflessly to provide the data. Importantly, we thank the growers and scientists of the Wild Blueberry Association of America for their support. In particular, we thank Susan Davis. We also thank Christa Thomas, Grace Millsap, Julie Macon Stegall, Kelly Sheppard, Kim Adams, Sheau Ching Chai, and Andrea Armer for their help with recruiting and data collection. This project was funded through USDA-ARS Project Nos. 0204-41510-001-24S, 58-6250-6-003, and R01DK092238. The contents of this publication do not necessarily reflect the views or policies of the US Department of Agriculture, nor does mention of trade names, commercial products, or organizations imply endorsement by the US Government.

Author Contributions: Carol L. Cheatham designed and conducted the larger study; V. Saroja Voruganti proposed the uric acid hypothesis; Itzel Vazquez-Vidal and Amanda Medlin managed the data and did the literature review; Carol L. Cheatham and V. Saroja Voruganti analyzed the data and wrote the paper; Carol L. Cheatham had primary responsibility for the final content. All authors read and approved the final manuscript.

Conflicts of Interest: The authors declare no conflict of interest.

Abbreviations

The following abbreviations are used in this manuscript:

BMI	body mass index
CKD	chronic kidney disease
CVD	cardiovascular disease
SUA	serum uric acid

References

1. Lippi, G.; Montagnana, M.; Franchini, M.; Favaloro, E.J.; Targher, G. The paradoxical relationship between serum uric acid and cardiovascular disease. *Clin. Chim. Acta* **2008**, *392*, 1–7. [CrossRef] [PubMed]

2. Ames, B.N.; Cathcart, R.; Schwiers, E.; Hochstein, P. Uric acid provides an antioxidant defense in humans against oxidant- and radical-caused aging and cancer: A hypothesis. *Proc. Natl. Acad. Sci. USA* **1981**, *78*, 6858–6862. [CrossRef] [PubMed]

3. Fabrinni, E.; Serafini, M.; Baric, I.C.; Hazen, S.L.; Klein, S. Effect of plasma uric acid on antioxidant capcaity, oxidative stress, and insulin sensitivity in obese subjects. *Diabetes* **2014**, *63*, 976–981. [CrossRef] [PubMed]

4. Sautin, Y.Y.; Nakagawa, T.; Zharikov, S.; Johnson, R.J. Adverse effects of the classic antioxidant uric acid in adipocytes: NADPH oxidase-mediated oxidative/nitrosative stress. *Am. J. Physiol. Cell. Physiol.* **2007**, *293*, C584–C596. [CrossRef] [PubMed]

5. Fang, J.; Alderman, M.H. Serum uric acid and cardiovascular mortality the NHANES I epidemiologic follow-up study, 1971–1992. National health and nutrition examination survey. *JAMA* **2000**, *283*, 2404–2410. [CrossRef] [PubMed]

6. Niskanen, L.K.; Laaksonen, D.E.; Nyyssonen, K.; Alfthan, G.; Lakka, H.M.; Lakka, T.A.; Salonen, J.T. Uric acid level as a risk factor for cardiovascular and all-cause mortality in middle-aged men: A prospective cohort study. *Arch. Intern. Med.* **2004**, *164*, 1546–1551. [CrossRef] [PubMed]

7. Culleton, B.F.; Larson, M.G.; Kannel, W.B.; Levy, D. Serum uric acid and risk for cardiovascular disease and death: The framingham heart study. *Ann. Intern. Med.* **1999**, *131*, 7–13. [CrossRef] [PubMed]

8. Arts, I.C.; Hollman, P.C. Polyphenols and disease risk in epidemiologic studies. *Am. J. Clin. Nutr.* **2005**, *81*, 317S–325S. [PubMed]

9. Scalbert, A.; Manach, C.; Morand, C.; Remesy, C.; Jimenez, L. Dietary polyphenols and the prevention of diseases. *Crit. Rev. Food Sci. Nutr.* **2005**, *45*, 287–306. [CrossRef] [PubMed]
10. Manach, C.; Mazur, A.; Scalbert, A. Polyphenols and prevention of cardiovascular diseases. *Curr. Opin. Lipidol.* **2005**, *16*, 77–84. [CrossRef] [PubMed]
11. Pandey, K.B.; Rizvi, S.I. Plant polyphenols as dietary antioxidants in human health and disease. *Oxid. Med. Cell. Longev.* **2009**, *2*, 270–278. [CrossRef] [PubMed]
12. Boban, M.; Modun, D.; Music, I.; Vukovic, J.; Brizic, I.; Salamunic, I.; Obad, A.; Palada, I.; Dujic, Z. Red wine induced modulation of vascular functioin: Separating the role of polyphenols, ethanol and urates. *J. Cardiovasc. Pharmacol.* **2006**, *47*, 695–701. [CrossRef] [PubMed]
13. Modun, D.; Music, I.; Vukovic, J.; Brizic, I.; Katalinic, V.; Obad, A.; Palada, I.; Dujic, Z.; Boban, M. The increase in human plasma antioxidant capacity after red wine consumption is due to both plasma urate and wine polyphenols. *Atherosclerosis* **2008**, *197*, 250–256. [CrossRef] [PubMed]
14. Jacob, R.A.; Spinozzi, G.M.; Simon, V.A.; Kelley, D.S.; Prior, R.L.; Hess-Pierce, B.; Kader, A.A. Consumption of cherries lowers plasma urate in healthy women. *J. Nutr.* **2003**, *133*, 1826–1829. [PubMed]
15. Sadowska-Krepa, E.; Klapcinska, B.; Podgorski, T.; Szade, B.; Tyl, K.; Hadzik, A. Effects of supplementation with acai (*Euterpe oleracea Mart.*) berry-based juice blend on the blood antioxidant defence capcacity and lipid profile in junior hurdlers. A pilot study. *Biol. Sport* **2015**, *32*, 161–168. [CrossRef] [PubMed]
16. Cassidy, A.; Mukamal, K.J.; Liu, L.; Franz, M.; Eliassen, A.H.; Rimm, E.B. High anthocyanin intake is associated with a reduced risk of myocardial infarction in young and middle-aged women. *Circulation* **2013**, *127*, 188–196. [CrossRef] [PubMed]
17. Cassidy, A.; O'Reilly, E.J.; Kay, C.; Sampson, L.; Franz, M.; Forman, J.P.; Curhan, G.; Rimm, E.B. Habitual intake of flavonoid subclasses and incident hypertension in adults. *Am. J. Clin. Nutr.* **2011**, *93*, 338–347. [CrossRef] [PubMed]
18. Wedick, N.M.; Pan, A.; Cassidy, A.; Rimm, E.B.; Sampson, L.; Rosner, B.; Willett, W.; Hu, F.B.; Sun, Q.; van Dam, R.M. Dietary flavonoid intakes and risk of type 2 diabetes in US men and women. *Am. J. Clin. Nutr.* **2012**, *95*, 925–933. [CrossRef] [PubMed]
19. Stull, A.J.; Cash, K.C.; Champagne, C.M.; Gupta, A.K.; Boston, R.; Beyl, R.A.; Johnson, W.D.; Cefalu, W.T. Blueberries improve endothelial function, but not blood pressure, in adults with metabolic syndrome: A randomized, double-blind, placebo-controlled clinical trial. *Nutrient* **2015**, *7*, 4107–4123. [CrossRef] [PubMed]
20. Johnson, S.A.; Figueroa, A.; Navaei, N.; Wong, A.; Kalfon, R.; Ormsbee, L.T.; Feresin, R.G.; Elam, M.L.; Hooshmand, S.; Payton, M.E.; et al. Daily blueberry consumption improves blood pressure and arterial stiffness in postmenopausal women with pre and stage1 hypertension: A randomized, double-blind, placebo-controlled clinical trial. *J. Acad. Nutr. Diet.* **2015**, *115*, 369–377. [CrossRef] [PubMed]
21. Rodriguez-Mateos, A.; Rendeiro, C.; Bergillos-Meca, T.; Tabatabaee, S.; George, T.W.; Heiss, C.; Spencer, J.P. Intake and time dependence of bluberry flavonoid-induced improvements in vascular function: A randolized, controlled, double-blind, crossover intervention study with mechanistic insights into biological activity. *Am. J. Clin. Nutr.* **2013**, *98*, 1179–1191. [CrossRef] [PubMed]
22. Blacker, B.C.; Snyder, S.M.; Eggett, D.L.; Parker, T.L. Consumption of blueberries with a high-carbohydrate, low-fat breakfast decreases postprandial serum markers of oxidation. *Br. J. Nutr.* **2013**, *109*, 1670–1677. [CrossRef] [PubMed]
23. Gustafson, S.J.; Yousef, G.G.; Grusak, M.A.; Lila, M.A. Effect of postharvest handling on phytochemical concentrations and bioactive potential in wild blueberry fruit. *J. Berry Res.* **2012**, *2*, 215–227.
24. Serafini, M.; Testa, M.F.; Villano, D.; Pecorari, M.; van Wieren, K.; Azzini, E.; Brambilla, A.; Maiani, G. Antioxidant activity of blueberry fruit is impaired by association with milk. *Free Radic. Biol. Med.* **2009**, *46*, 769–774. [CrossRef] [PubMed]
25. Wu, A.H.; Gladden, J.D.; Ahmed, M.; Ahmed, A.; Filippatos, G. Relation of serum uric acid to cardiovascular disease. *Int. J. Cardiol.* **2016**, *213*, 4–7. [CrossRef] [PubMed]
26. Voruganti, V.S.; Franceschini, N.; Haack, K.; Laston, S.; MacCluer, J.W.; Umans, J.G.; Comuzzie, A.G.; North, K.E.; Cole, S.A. Replication of the effect of SLC2A9 genetic variation on serum uric acid levels in American Indians. *Eur. J. Hum. Genet.* **2014**, *22*, 938–943. [CrossRef] [PubMed]

27. Voruganti, V.S.; Laston, S.; Haack, K.; Mehta, N.R.; Cole, S.A.; Butte, N.F.; Comuzzie, A.G. Serum uric acid concentrations and SLC2A9 genetic variation in Hispanic children: The Viva La Familia Study. *Am. J. Clin. Nutr.* **2015**, *101*, 725–732. [CrossRef] [PubMed]

28. Voruganti, V.S.; Nath, S.D.; Cole, S.A.; Thameem, F.; Jowett, J.B.; Bauer, R.; MacCluer, J.W.; Blangero, J.; Comuzzie, A.G.; Abboud, H.E.; et al. Genetics of variation in serum uric acid and cardiovascular risk factors in Mexican Americans. *J. Clin. Endocrinol. Metab.* **2009**, *94*, 632–638. [CrossRef] [PubMed]

29. Liu, M.; He, Y.; Jiang, B.; Wu, L.; Yang, S.; Wang, Y.; Li, X. Association between serum uric acid level and metabolic syndrome and its sex difference in a Chinese community elderly population. *Int. J. Endocrinol.* **2014**, *2014*, 754678. [CrossRef] [PubMed]

30. MacCluer, J.W.; Scavini, M.; Shah, V.O.; Cole, S.A.; Laston, S.L.; Voruganti, V.S.; Paine, S.S.; Eaton, A.J.; Comuzzie, A.G.; Tentori, F.; et al. Heritability of measures of kidney disease among Zuni Indians: The Zuni Kidney Project. *Am. J. Kidney Dis.* **2010**, *56*, 289–302. [CrossRef] [PubMed]

31. Zurlo, A.; Veronese, N.; Giantin, V.; Maselli, M.; Zambon, S.; Maggi, S.; Musacchio, E.; Toffanello, E.D.; Sartori, L.; Perissinotto, E.; et al. High serum uric acid levels increase the risk of metabolic syndrome in elderly women: The PRO. VA study. *Nutr. Metab. Cardiovasc. Dis.* **2016**, *26*, 27–35. [CrossRef] [PubMed]

32. Moulin, S.R.; Baldo, M.P.; Souza, J.B.; Luchi, W.M.; Capingana, D.P.; Magalhaes, P.; Mill, J.G. Distribution of serum uric acid in black africans and its association with cardiovascular risk factors. *J. Clin. Hypertens.* **2016**. [CrossRef] [PubMed]

33. Kawabe, M.; Sato, A.; Hoshi, T.; Sakai, S.; Hiraya, D.; Watabe, H.; Kakefuda, Y.; Ishibashi, M.; Abe, D.; Takeyasu, N.; et al. Gender differences in the association between serum uric acid and prognosis in patients with acute coronary syndrome. *J. Cardiol.* **2016**, *67*, 170–176. [CrossRef] [PubMed]

34. Chou, P.; Lin, K.C.; Lin, H.Y.; Tsai, S.T. Gender differences in the relationships of serum uric acid with fasting serum insulin and plasma glucose in patients without diabetes. *J. Rheumatol.* **2001**, *28*, 571–576. [PubMed]

35. Lippi, G.; Montagnana, M.; Luca Salvagno, G.; Targher, G.; Cesare Guidi, G. Epidemiological association between uric acid concentration in plasma, lipoprotein(a), and the traditional lipid profile. *Clin. Cardiol.* **2010**, *33*, E76–E80. [CrossRef] [PubMed]

36. Rodrigues, S.L.; Baldo, M.P.; Capingana, P.; Magalhaes, P.; Dantas, E.M.; Molina Mdel, C.; Salaroli, L.B.; Morelato, R.L.; Mill, J.G. Gender distribution of serum uric acid and cardiovascular risk factors: Population based study. *Arq. Bras. Cardiol.* **2012**, *98*, 13–21. [CrossRef] [PubMed]

37. Wang, S.F.; Shu, L.; Wang, S.; Wang, X.Q.; Mu, M.; Hu, C.Q.; Liu, K.Y.; Zhao, Q.H.; Hu, A.L.; Bo, Q.L.; et al. Gender difference in the association of hyperuricemia with hypertension in a middle-aged Chinese population. *Blood Press.* **2014**, *23*, 339–344. [CrossRef] [PubMed]

antioxidants

MDPI

Review

Protective Role of Dietary Berries in Cancer

Aleksandra S. Kristo [1,*], Dorothy Klimis-Zacas [2] and Angelos K. Sikalidis [1]

[1] Department of Nutrition and Dietetics, Istanbul Yeni Yuzyil University, Yilanli Ayasma Caddesi No. 26, Istanbul 34010, Turkey; as545@cornell.edu

[2] School of Food and Agriculture, University of Maine, Orono, ME 04469, USA; Dorothy_Klimis_Zacas@umit.maine.edu

* Correspondence: aleksandra.kristo@yeniyuzyil.edu.tr; Tel.: +90-212-444-5001; Fax: +90-212-481-4058

Academic Editor: Stanley Omaye
Received: 30 June 2016; Accepted: 11 October 2016; Published: 19 October 2016

Abstract: Dietary patterns, including regular consumption of particular foods such as berries as well as bioactive compounds, may confer specific molecular and cellular protection in addition to the overall epidemiologically observed benefits of plant food consumption (lower rates of obesity and chronic disease risk), further enhancing health. Mounting evidence reports a variety of health benefits of berry fruits that are usually attributed to their non-nutritive bioactive compounds, mainly phenolic substances such as flavonoids or anthocyanins. Although it is still unclear which particular constituents are responsible for the extended health benefits, it appears that whole berry consumption generally confers some anti-oxidant and anti-inflammatory protection to humans and animals. With regards to cancer, studies have reported beneficial effects of berries or their constituents including attenuation of inflammation, inhibition of angiogenesis, protection from DNA damage, as well as effects on apoptosis or proliferation rates of malignant cells. Berries extend effects on the proliferation rates of both premalignant and malignant cells. Their effect on premalignant cells is important for their ability to cause premalignant lesions to regress both in animals and in humans. The present review focuses primarily on in vivo and human dietary studies of various berry fruits and discusses whether regular dietary intake of berries can prevent cancer initiation and delay progression in humans or ameliorate patients' cancer status.

Keywords: antioxidants; anthocyanins; cancer; chemoprevention; edible berries; flavonoids; phytochemicals

1. Background

Cancer, the uncontrolled growth of cells which can invade and spread to distant sites of the body, is a global health problem with high mortality and disability rates. The most common forms of cancer in males are: lung, prostate, colorectal, stomach, and liver cancer; whereas in females: breast, colorectal, lung, uterine/cervix, and stomach cancer. Prevention, in coordination with monitoring leading to early and accurate diagnosis when a case is confirmed, is critical, since most therapeutic options do not confer cure but rather a deceleration of cancer progression aiming at life extension and improvement of patients' life quality, albeit with serious and often debilitating side-effects. According to the latest available World Health Organization (WHO) global report (World Cancer Report 2014, as updated in 2015) [1] cancer is one of the leading causes of morbidity and mortality worldwide, reaching approximately 14 million new cases and 8.2 million cancer related deaths in 2012 [1] (Table 1). While new cancer incidence is expected to rise by 70% by the year 2034, approximately 35% of cancer deaths are attributed to the five leading behavioral and dietary risks: high body mass index (BMI), low fruit and vegetable intake, lack of physical activity, tobacco use (primarily smoking) and harmful alcohol use [1].

Table 1. Deaths by major types of cancer in 2012 (world).

Type of Cancer	Deaths (Figures in Millions)
Lung	1.590
Liver	0.745
Stomach	0.723
Colorectal	0.694
Breast	0.521
Esophageal	0.400
Total	4.673

Source of data: WHO—world cancer report 2014.

According to the latest published report by the Center for Disease Control (CDC), in the US alone 1,685,210 new cases of cancer are estimated in 2016 of which 841,390 pertain to males and 843,820 to females. The same report projects an estimated 595,690 deaths from cancer in 2016 in the entire US, of which 314,290 refer to males and 281,400 to females [2]. According to the Agency for Healthcare Research and Quality (AHRQ) in the USA (all 50 States) the direct medical costs (total of all health care costs) for cancer in 2011 reached US $88.7 billion [2–4]. In its latest global report on cancer issued in 2014 as updated in 2015 [1], WHO stresses that being overweight or obese on one hand and consuming unhealthy diets with low fruit and vegetable intakes on the other, constitute important, yet modifiable risk factors which significantly increase cancer risk for both genders. A plethora of in vitro, in vivo as well as human studies has suggested diet as a critical factor for reducing cancer risk with a particular emphasis on certain foods/food groups, to which specific potential for cancer risk reduction has been attributed [1]. Diets rich in fruits and vegetables have been associated with a reduced risk of cancer [5–7]. Many non-nutrient plant compounds known as phytochemicals, including numerous phenolic compounds, have been identified, and found to exert anti-cancer activity [5,8]. In this context, there has been an increasing interest in the role of polyphenols and other phytochemicals in cancer, particularly for their anti-oxidant and anti-inflammatory properties. The intake of bioactive compounds such as flavonoids has been inversely correlated with systemic inflammatory markers in human populations [9,10]. Protective role of certain foods against cancer, while clearly suggested, cannot easily be delineated as food items are highly complex, whereas protective compounds naturally found in foods, arguably act synergistically when present in an optimal balance. Therefore, research on the effects of whole food consumption and/or dietary schemes in relation to cancer risk is being increasingly conducted as deemed more physiologically and practically significant.

2. Cancer Development and Associated Mechanisms

Cancer can be viewed as a gradual generation and development of a tumor, which from a functional perspective can be divided into three phases: initiation, promotion and progression. Genomic changes such as point mutations, gene deletion and amplification and chromosomal rearrangements, all mark initiation, subsequently committing the cell to an irreversible status. In order for a tumor to be formed however, survival and clonal expansion of these initiated cells is required (Figure 1). Growth of tumor and metastasis (if it occurs) are characteristics of the progression phase. Development of cancer is a multistage process involving multiple genetic and epigenetic events occurring at varying rates. In order for cancer to develop, the acquisition of all six properties below is required: (i) self-sufficient proliferation; (ii) insensitivity to anti-proliferative signals; (iii) evasion of apoptosis/T-cell control; (iv) unlimited cellular replication ability; (v) maintenance of vascularization; and (vi) tissue invasion and metastasis. These alterations result from a combination of proto-oncogene activation, inactivation of tumor suppressor genes and inactivation of genomic stability genes.

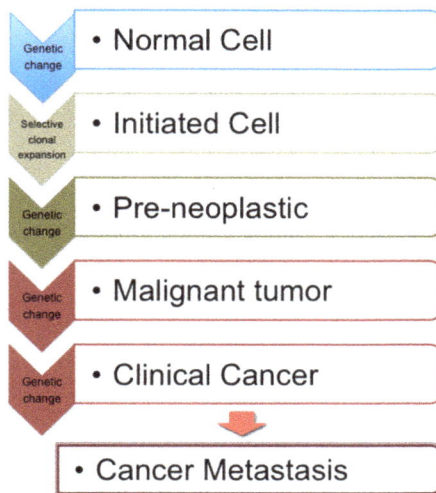

Figure 1. Schematic depiction of cancer development/progression.

Several conditions and/or mechanisms that function as risk modulators for cancer including oxidative stress/chronic inflammation, obesity/metabolic syndrome, angiogenesis, apoptosis, autophagy and proliferation, have been proposed to explain cancer. Depending on the type/case of cancer any of the above and/or a combination may lead to cancer manifestation or progression (Figure 2). There is however no definitive mode of action via which cancer is initiated and hence prediction of cancer occurrence is not possible, while from a statistical standpoint cancer prevention is possible to a certain degree.

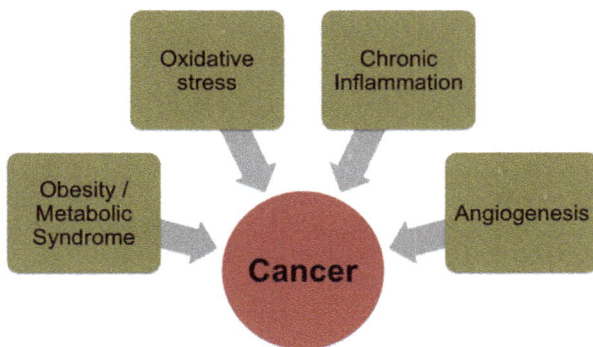

Figure 2. Major nutrition/diet-related factors/stressors that can contribute to cancer.

As cancer is a step-wise procedure, several factors may influence its initiation, development and progression. Oxidative stress can activate a variety of transcription factors including nuclear factor kappa B (NF-κB), activator protein-1 (AP-1), p53, hypoxia-inducible factor-1α (HIF-1α), peroxisome proliferator-activated receptor gamma (PPAR-γ or PPARG), β-catenin/Wnt, and NF-E2 related factor-2 (Nrf2). Activation of these transcription factors can lead to the expression of over 500 different genes, including those for growth factors, inflammatory cytokines, chemokines, cell cycle regulatory molecules, and anti-inflammatory molecules [11]. Chronic oxidative stress and inflammation can induce the events that essentially tip the balance so as to shift the status of a cell from the healthy to the malignant phenotype thus giving rise to tumorigenesis (Figure 3).

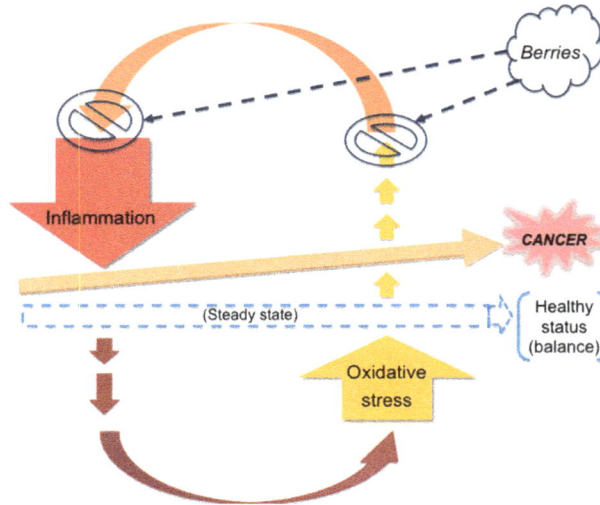

Figure 3. Conceptual schematic representation depicting the contribution of oxidative stress and inflammation in the deregulation/disturbance of homeostatic balance in a cell, thus leading to cancer. Berries by means of their numerous constituents (synergistic action) can interrupt this vicious cycle, thereby extending protective effects against cancer.

Biological, chemical, and physical factors can all, either in combination or separately, lead to oxidative stress resulting in the production of reactive oxygen species (ROS) and/or reactive nitrogen species (RNS). Oxidative stress causes damage to tissues and elicits an immune response that induces inflammation in an attempt of the body to rectify the inflicted damage. During the inflammatory response, mast cells and leukocytes are recruited to the damaged site, leading to a "respiratory burst" because of an increased requirement and thus uptake of oxygen, hence increasing release and accumulation of ROS at the site of damage, a development further perpetuating the damage to the already damaged and inflamed site. As the level of inflammatory response is augmented, a series of signaling molecules are produced by the inflammatory cells such as metabolites of arachidonic acid, cytokines, and chemokines, acting to further intensify the inflammatory responses and the continuing production and accumulation of ROS.

Hence, oxidative stress increases in the microenvironment of the damaged site. Simultaneously, the signaling agents produced induce signaling cascades involving NF-κB, signal transducer and activator of transcription 3 (STAT3), HIF1-α, AP-1, nuclear factor of activated T-cells (NFAT) and Nrf2, which mediate immediate cellular stress responses. This includes induction of cyclooxygenase-2 (COX-2), inducible nitric oxide synthase (iNOS), aberrant expression of inflammatory cytokines: tumor necrosis factor α (TNF-α), interleukin-1 (IL-1), IL-6 and chemokines such as IL-8; CXC chemokine receptor 4 (CXCR4). This microenvironment now characterized by increasingly high levels of inflammation and oxidative stress can spread damage to neighboring cells/tissues and, if such condition is sustained and becomes chronic, it can eventually lead to carcinogenesis. Therefore, chronic inflammation, as mediated by the induction of the immune system due to oxidative stress, can predispose the host to cancer among various chronic illnesses [11–16].

Apart from the oxidative stress/inflammation axis and the related mechanism for induction of carcinogenesis, tumor cells activate autophagy in response to cellular stress and/or increased metabolic demands related to rapid cell proliferation [17]. Autophagy-related stress tolerance can induce cellular pro-survival mechanisms thus leading to tumor growth and therapeutic resistance. Interestingly, work with preclinical models, demonstrated that inhibition of autophagy can restore chemosensitivity and enhance tumor cell death thus improving response to therapy [17].

3. Risk Factors for Cancer

Although cancer is not exclusively a disease of the elderly, age constitutes one of the risk factors for cancer, as the majority of people diagnosed with the disease are 65 years or older. Certain habits/lifestyle choices are shown (epidemiologically) to increase risk of cancer, including use of tobacco and tobacco products (particularly smoking), drinking more than one alcoholic drink a day (for women of all ages and men older than age 65) or two drinks a day (for men age 65 and younger), excessive exposure to the sun or frequent blistering sunburns, and obesity [1,4]. A small, yet considerable proportion of cancers are due to an inherited condition. Although inherited genetic mutations do not necessarily lead to cancer development, cancer risk is increased, as the probability is significantly higher. The environment (natural and anthropogenic) may expose an individual to harmful substances, such as air pollutants, water pollutants, and various hazardous chemicals and other toxicants. Examples of chemical carcinogens are asbestos and benzene, and they are associated with an increased risk of cancer primarily of the lungs [1,4]. Exposure to carcinogens increases the risk of cancer at different degrees according to the type of carcinogen as the carcinogenicity/toxicity levels of compounds vary [18]. Infectious agents, such as viruses and bacteria increase cancer risk (e.g., human papillomavirus infections and higher risk of cervical cancer in women). Radiation, including ionizing radiation and non-ionizing radiation constitute risk factors for a variety of cancers [18]. Additional cancer risk factors include pharmaceutical agents and exogenous and endogenous hormones. Moreover, poor immune system status and inflammation, particularly chronic inflammation, are positively associated with cancer risk (e.g., *Helicobacter pylori* infection is strongly related to increased gastric cancer risk) [18].

Overweight/obese status is associated with an increased risk for many cancer types such as postmenopausal breast cancer, endometrial cancer, colorectal, esophageal, gallbladder, kidney, pancreatic and thyroid cancer [18]. Most of the molecules that are being investigated as potential mediators between obesity and cancer are cancer-promoting rather that cancer-causing per se, hence they do not interfere with DNA damage/correction and mutations; instead they induce growth and proliferation of malignant cells, while others are also involved in promoting metastasis [18]. However, the association between obesity and cancer is strong. According to data presented by National Cancer Institute (NCI) and the International Agency for Research on Cancer using European data, in 2002 alone 39% of endometrial cancers, 37% of esophageal cancers and 25% of kidney cancers were related to obesity. Moreover, the American Cancer Society, as reported in 2003, attributes 14% of all cancers in males and 20% of all cancers in females to excess weight [18]. Interestingly, on a population level, the overweight/obesity-attributable number of cancers is approximately equal to that attributable to current smoking. This surprising observation may be explained partly by the decreasing trends in smoking and simultaneously increasing trends for obesity. On an individual basis nevertheless, cancer risk due to smoking remains substantially higher than that attributed to obesity [18]. As per smoking, apart from first-hand smoking, second- and third-hand smoking also appear to contribute to increased risk for cancer, yet at a much lower rate [18].

A key environmental factor that interacts with the human organism at several levels is food/diet. Dietary constituents individually, synergistically and/or as a whole can influence the way the human body will respond from a biochemical perspective (i.e., metabolically), from a signaling perspective and from an epigenetic perspective among others. In this context, diet can be used as a tool to evoke the positive/desirable biological responses of an organism aiming to maximize health and protection against diseases (particularly chronic/non-communicable diseases) by mostly means of prevention.

Certain food groups such as fruits and vegetables among others have been significantly studied in this regard. Fruits and vegetables in particular have been suggested to exert cancer protective effects due to different mechanisms such as inhibition of carcinogen activation, stimulation of carcinogen detoxification, scavenging of free radical species, control of cell-cycle progression, inhibition of cell proliferation, induction of apoptosis, inhibition of the activity of oncogenes, inhibition of angiogenesis and metastasis, and inhibition of hormone or growth-factor activity [5,19].

4. Berries and Cancer

4.1. Rationale of the Current Review

Edible berries are being increasingly investigated for their potential to extend chemoprevention as well as protection against a variety of chronic diseases. Here we discuss the protective role of dietary berries in cancer. The focus of this review, regarding the selection of studies, was predominantly placed on in vivo and human studies. There is certainly an abundance of in vitro studies in the scientific literature where cell-lines and berry extracts are used to delineate the chemopreventive effects of berry extracts and/or specific compounds naturally found in berries at significant amounts. Even though these studies are indicative of the potential berries have in exerting favorable effects in cases of cancer, the systems are very specific and to a large extent notably simple and isolated from the physiological state. Therefore, these studies are often met with the criticism that they are limited in terms of their physiological significance and consequently exhibit limited applicability and translational potential. The reductionist approach and mechanistic studies along with logical reasoning have proposed some potentially key players in terms of chemoprevention and/or improvement of cancer prognosis in berries, however these compounds in isolation do not seem to extend the promised and desired effects optimally. It is highly uncertain which particular compounds extend the overall observed effects in humans. Most likely it is the synergistic action at varying degrees of contribution of a plethora of phytochemicals/bioactives complexly intervening at several molecular pathways. Therefore, from a nutritionist's standpoint whole berry consumption is strongly advocated.

In this line of thought, our decision was to review and discuss the studies that were performed, either with intact animals or were done with humans, and where the form of the berries used was natural whole food. In vivo studies using freeze-dried berries are included as well. In terms of the berry types, our discussion was led by the availability of in vivo and human studies that used whole berries irrespective of what the berry itself was. We do present a limited number of characteristics from in vitro works that show bioactivity and chemoprevention potential for typical berry-found compounds as an indicator of potential mechanistic evidence for the mode of action towards chemoprevention and a lead-in to the core discussion.

A search for literature on in vivo and human studies investigating the effect of berry consumption on cancer initiation, progression, metastasis and overall risk was conducted. Relevant abstracts and full texts were screened. PubMed, Science Direct, Web of Science and Google Scholar databases were searched to identify articles published later than 1 January 2000. The searches used the following terms and text words alone and in combinations: "berry/berries", "cancer", "chemoprevention", "cancer risk", "anthocyanins", "flavonoids", "polyphenols", "phytochemicals", "animals", and 'humans'. Reference lists of the obtained articles were also searched for additional articles. The research was limited to English-language written articles. A total of 61 relevant articles were obtained from the database searches and from the reference lists of the obtained papers that met the aforementioned criteria. These studies were included in the review along with supportive data obtained from characteristic in vitro studies as well as evidence provided and discussed by recent review manuscripts. Additionally, we reviewed data from recent official reports issued by the World Health Organization, Center for Disease Control, and the European Commission. Here, we summarize the main results and conclusions, and discuss the findings of in vivo and human studies testing berries in terms of their anti-cancer properties.

4.2. Berry Types and Composition

According to strict botanical terminology a "berry" is a simple fruit with seeds and pulp produced from the ovary of a single flower and the pericarp (fruit wall) is fleshy (i.e., the fruit is fleshy from the skin layer inwards and all throughout to its core but the seeds) [20]. Under common usage however, the term berry refers to a small pulpy and often edible fruit. For example, blueberries and cranberries would be categorized as berries under both definitions, while bananas, tomatoes, grapes and pumpkin are berries according to the botanical definition, whereas blackberries, raspberries and strawberries, even though they are typically considered and referred to as berries by the lay audience, from a strictly botanical perspective they are not categorized as such. Hence, the term "berry" can indeed be fairly confusing since there are occasional discrepancies between the strict botanical criteria set, and what is widely used and/or accepted as being a berry. The studies included in this review refer mostly to fruits commonly consumed and recognized as berries.

Overall, berries are rich in polyphenols and most anthocyanins. Anthocyanins, along with other flavonoids, are localized in the skin, seeds and leaves of the berries giving them their distinctive pigmentation. Polyphenols constitute the largest group of phytochemicals found in plants, particularly in fruits, seeds, and leaves. There are more than 8000 chemical compounds found in the human diet today that are identified as dietary polyphenols. They are secondary metabolites of plants that contain one or more hydroxyl (–OH) group(s) attached to -ortho, -meta or -para position(s) on a typical benzene ring. These metabolites are generally involved in defense against ultraviolet radiation, various environmental pollutants, and pathogens [21]. Flavonoids on the other hand, constitute a very large (more than 6000 have been identified thus far) and very diverse group of phytonutrients (Figures 4 and 5). Currently, even though a surplus of data from in vitro studies suggests antioxidant benefits from berries (particularly berry-extracts) due to several bioactive compounds—mostly polyphenols and flavonoids, (see Supplementary Materials Table S1)—a clear consensus stemming from in vivo and corroborated from human studies is missing. Health claims that foods containing polyphenols have antioxidant health value for consumers are not permitted on product labels by the official regulatory authorities in the USA (Food and Drug Administration, Silver Spring, MD, USA) or the European Union (EU-28), (European Food Safety Organization, Parma, Italy). Nonetheless, berries are widely considered functional foods that may well extend several health benefits.

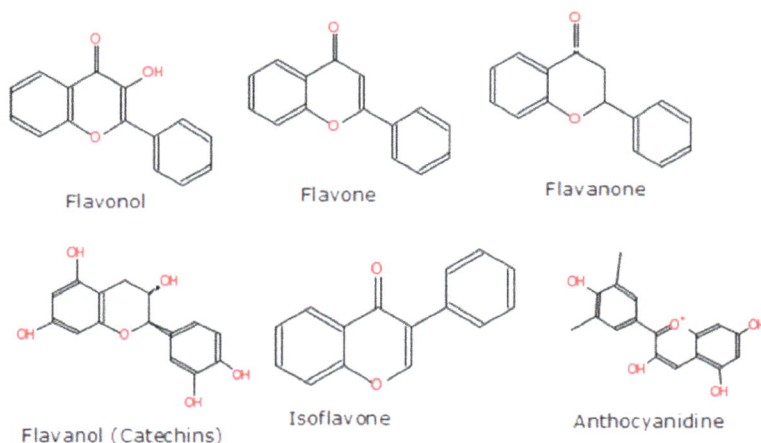

Figure 4. Major types of flavonoids (chemical structures produced via the eMolecules platform developed by eMolecules Inc., La Jolla, CA, USA).

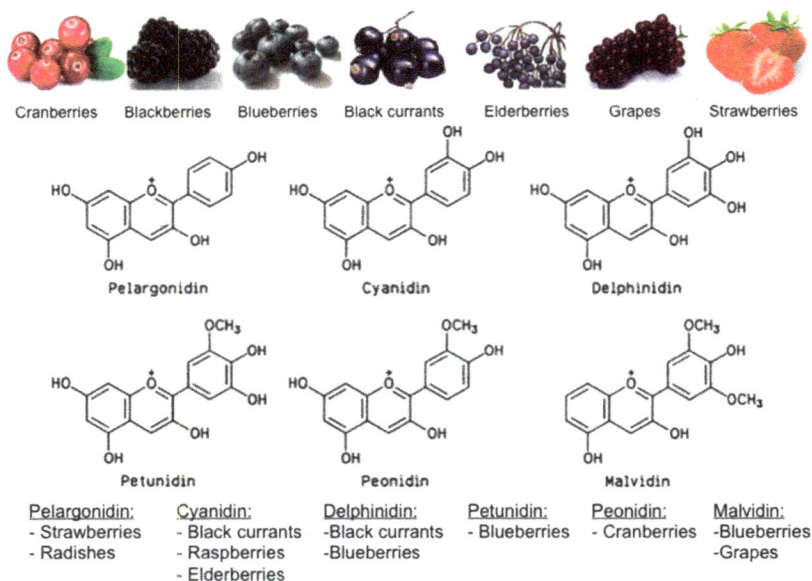

Figure 5. Major types of anthocyanins and characteristic examples of foods (berries) rich in the respective compounds (chemical structures produced via the eMolecules platform developed by eMolecules Inc., La Jolla, CA, USA).

4.3. Mechanisms Associated with Berries' Anticancer Capacity

Edible berries are receiving increasing attention due to the great variety of phytochemicals, including but not limited to typical antioxidants, linked to protection against cancer among other chronic diseases. Berry constituents that have been suggested to exert cancer protective effects in cells include: phenolic acids (hydroxycynnamic acid, hydroxybenzoic acid), stilbenes (resveratrol, pterostilbene, piceatannol), flavonoids (anthocyanins, flavonols, catechins), lignans, tannins (proanthocyanidins, ellagitannins).

Berries are notably rich in compounds that are shown to exert potential for chemoprevention and particularly flavopiridol, ellagic acid, anethole and resveratrol have been demonstrated to inhibit the NF-κB signaling pathway, either at the point of signaling cascade activation or at the point of NF-κB's translocation into the nucleus. Other points of interference include the DNA binding dimmers and/or interactions with the basal transcriptional machinery [22]. Resveratrol and anethole were also found to inhibit AP-1, which is linked to growth regulation and cell transformation. Further, AP-1 seems to be involved in the regulation of genes involved in apoptosis and proliferation while it can promote the transition of a tumor cell from epithelial to mesenchymal morphology, an early step marking metastasis [22]. Resveratrol and flavopiridol were demonstrated to down-regulate the expression of apoptosis suppressor proteins (Bcl-2 and Bcl-XL) in a variety of cancer cell-lines [22]. A variety of phytochemicals such as indole-3-carbinol, curcuminoids and epigallocatechin-3-gallate (EGCG) (as well as black raspberries as a whole fruit) have been shown to suppress Akt activation thus leading to cancer suppression signaling. Moreover, several common constituents of berries such as quercetin, kaempferol and pterostilbene were found to attenuate ROS in HepG2-C8 cells via the Nrf2-Antioxidant Response Element signaling pathway, suggesting that induction of antioxidant defense is likely one of the mechanisms via which berries provide chemoprevention [23]. Polyphenols are also likely to modify the redox state of cancer microenvironment thus rendering it more cytotoxic hence leading to apoptosis secondary to induced oxidative stress [24,25]. In various systems, berries have been shown to down-regulate all major inflammatory markers such as TNF-α, IL-1β, IL-6, IL-10, iNOS, COX-2, PGE$_2$, NF-κB and p-p65 [24–27]. For example, anti-inflammatory pathways that have been shown induced by berry derived anthocyanin rich extracts include the reduction of expression levels of iNOS, COX-2, IL-1β and IL-6 in RAW264.7 macrophages [28]. In this regard, inflammation suppression may lead to protection against cancer occurrence and/or progression. The degree to which such protection is extended is yet unclear but appears significant. Additionally, berries are demonstrated to reduce cellular proliferation by down-regulating PCNA and Ki-67/MKI67, as well as inhibiting signaling pathways such as the PI3K/Akt/mTOR axis, MAPK/ERK and Wnt pathways [24]. Furthermore, anti-cancer effects can be extended via the inhibition of STAT3 and cell cycle arrest, thus impeding proliferation of cancer cells. Hence, cancer inhibition of cellular growth and proliferation may function towards deceleration of cancer progression thus improving the possibility for anti-cancer treatment success and subsequent patient's survival. Another potential mechanism via which berries were shown to confer protection against cancer development is the induction of apoptosis. Berries may confer chemoprevention via the activation of caspases and the mitochondrial damage/cytochrome c pathways hence leading to enhanced apoptosis of cancer cells [23]. Berries were observed to up-regulate p53, caspases-3, -8 and -9 as well as inducing Bax and Cyto-c while down-regulating Bcl-2 and PARP [24,25]. Therefore the progression of cancer can be decelerated thus assisting the treatment efforts, consequently improving survival rates of patients. Cell cycle regulation also appears to constitute yet another potential mechanism through which berries may exert their anticancer function. More specifically, berries were found to increase p16, p21 and p27 while reduce a series of cell cycle markers such as CDK$_2$, CDK$_4$, cyclin A, cyclin B1 and cyclin D1, Cdc2, Cdc25C and the Rb protein [24]. Such synergies make a cell less prone to cancerous transformation and reduce the potential for tumorigenesis. The reduction of angiogenesis has also been proposed as a potential mechanism of berries' function. A series of pro-angiogenic factors such as c-Myc, c-jun, c-fos and VEGF, have been shown markedly reduced by berries [24]. Cell adhesion proteins such as β-catenin, ICAM-1 and VCAM-1 are attenuated by berries [24], while cellular invasion is also inhibited through the repression of MMP and u-PA by anthocyanins commonly found in berries [29]. These observations underline cell adhesion attenuation and cellular invasion suppression as additional potential mechanisms via which berries may provide anti-cancer protection. As for cell adhesion, invasion and migration along with angiogenesis are all phenomena that facilitate metastasis, berries may extend inhibition of metastasis by suppressing such pro-metastatic phenomena. Additionally, effects on phase-I and -II enzymes have been proposed as potential mechanisms of berries' anti-cancer function [29]. More specifically, polyphenols have been shown to inhibit strongly phase-I detoxifying enzyme CYP1A1 while inducing phase-II detoxifying enzymes GST and NADPH quinone oxidoreductase (NQO) in mouse epidermal cells [29].

Furthermore, other modes of berry activity via certain bioactive compounds they contain involve modulation of miRNA expression profiles, DNA methylation and histone modifications, all of which can lead to the inhibition of cancer cell growth, induction of apoptosis, reversal of epithelial-mesenchymal transition, or improvement of conventional cancer therapeutics' efficacy [30]. Epigenetic alterations such as inhibition of histone deacetylases (HDACs), miRNAs and modification of the CpG methylation of cancer-related genes may indeed be key mechanisms by which berries reduce cancer risk, especially when taking into account the significantly lower concentrations of bioactive compounds in human blood compared to in vitro testing [23,30]. A plausible explanation for their activity in humans, even at markedly lower concentrations, could be that these compounds exert their biological activities through epigenetic modulation. There is growing interest in dietary compounds that confer favorable epigenetic modulation against chronic diseases such as cancer. Characteristic is the example of resveratrol, which activates class III HDACs (sirtuins) because of their potential role in extending lifespan and in reducing, or delaying, age-related diseases including cancers [30]. A summary of the mechanisms through which berries may extend chemoprevention and/or evoke therapeutic responses post-cancer occurrence is provided in Table 2. Below we discuss evidence from studies indicating anti-cancer function of specific bioactive compounds typically found in berries.

Table 2. Summary table of mechanisms and mode of action via which berries may evoke chemopreventive and therapeutic responses against cancer.

Mechanism	Mode of Action	Bioactive Compound *
Anti-oxidative action	ROS sequestration ↑ GSH	Epigallocatechin-3-gallate (EGCG)
Effects on enzymes of phase-I and -II	↓ CYP1A1 ↓ LDH	Quercetin, kaempferol
	↑ UDPGT ↑ NQO	Ellagic acid Chlorogenic acid
Cell-cycle arrest	↓ cyclin D, E ↓ CDK 1, 2, 4 ↓ PCNA	EGCG
	↑ cyclin E ↓ cyclin A, B1	Ellagic acid
Apoptosis	↑ ROS in cancer cells ↑ caspase-3, -7, -8 ↑ cytochrome c ↑ Blc-2, Blc-X_L	EGCG
	↑ caspase-3, -7, -9 ↑ cytochrome c ↑ PARP cleavage	Quercetin
Anti-proliferation/Anti-survival	↓ GFR/Ras/MAPK & PI3K/Akt ↓ c-fos ↓ erg-1 ↓ PI3K ↓ ERK ↓ Akt phosphorylation ↓ NF-κB	EGCG
Anti-inflammatory action	↓ COX-1 ↓ COX-2	Gallic acid
	↓ TNF-α ↓ COX-2	EGCG
Anti-angiogenesis	↓ VEGF ↓ PDGF ↓ HIF-1α	EGCG, anthocyanin berry extracts
Metastasis inhibition	↓ MMP-9 ↓ mRNA stabilizing factor HuR	EGCG
Cell adhesion and movement inhibition	↓ MRLC phosphorylation ↓ Vimentin phosphorylation	EGCG

Abbreviations: Akt/PKB, Protein kinase B; CDK, Cyclins-dependent kinase; COX, Cyclooxygenase; CYP, Cytochrome P450; ERK, Extracellular regulated kinase; GFR, Growth factor receptors; GSH, Glutathione; HIF-1α, Hypoxia-inducible factor 1α; LDH, Lactate dehydrogenase; MAPK, Mitogen-activated protein kinase; MMP-9, Matrix metallopeptidase 9; MRLC, Myosin II regulatory light chain (protein); NF-κB, Nuclear factor—kappa (κ) B; NQO, NADPH quinone oxidoreductase; PARP, Poly-ADP ribose polymerase; PCNA, Proliferating cell nuclear antigen; PDGF, Platelete-derived growth factor; PI3K, Phosphatidylinositol-3-kinase; ROS, Reactive oxygen species; TNF-α, Tumor necrosis factor alpha (α); UDPGT, UDP-glucuronosyl transferase; VEGF, Vascular endothelial growth factorr. ↑ Up-vector indicates induction/increase. ↓ Down-vector indicates suppression/decrease. * Other berry-derived bioactive compounds may confer similar response.

4.4. Berries' Potential for Cancer Risk Reduction Due to Specific Constituents

A plethora of berry-derived compounds has been studied for their chemopreventive properties [31], but also as a proxy for establishing potential mechanisms via which berries may be extending chemoprevention as suggested by epidemiology. The major phenolics isolated from the apple and berry juice, such as phlorizin, rutin, quercetin and their two glucoside forms, and phloretin, exhibited significant inhibition of cytochrome's P4501A1 enzymatic activity [32]. Kern and colleagues showed that the activity of protein kinase C (PKC), a well-established signaling molecule involved in carcinogenesis induction, is significantly reduced in HT29 cells when incubated with polyphenol-rich apple juice extract (AE02) for 24 h although the result was not sustained. More interestingly, prolonged incubation of HT29 cells with AE02 resulted in significantly induced apoptosis via the activation of caspase-3, DNA fragmentation, and cleavage of poly(ADP ribose) polymerase [26,33]. Wang and Jiao [26] showed that ROS such as super-oxide radical and hydrogen peroxide (H_2O_2), were found to be sequestered by different berry-juices (i.e., blackberry, strawberry, raspberry, cranberry and blueberry), thus offering a potential mechanism for cancer risk reduction. Besides the direct scavenging ability of berries, their antioxidant potential due to the polyphenols they contain is also an element that further enhances their anti-cancer properties. Additionally, antioxidant enzymes such as catalase, glutathione reductase and ascorbate peroxidase were identified in significant concentrations in strawberries and blackberries thus offering yet another potential means of cancer chemoprevention extended by berries. The combination of anti-ROS elements in berries could mount a resistance at the very first level/step of carcinogenesis; ROS-induced DNA damage. On the other hand, DNA repair seems to be promoted through the induction of oxidative adducts' removal from the DNA at least in some cases. In a study with mice, administration of ferulic acid, a phenolic acid commonly found in berries, was demonstrated to significantly reduce DNA strand breaks after whole body γ-irradiation [34]. Berries (i.e., strawberries, black raspberries and blackberries) in their lyophilized form as part of diet in rats were shown to significantly inhibit the appearance and development of carcinogen-induced tumors [35].

In a review of the evidence comprised by Lall and colleagues, the relationship between polyphenols and prostate cancer was examined. Although the results are occasionally inconsistent and somewhat variable, there is a consensus that polyphenols constitute promising agents for the management of prostate cancer. Gallic acid (GA) for example has been shown to exert anti-cancer function in human PCa DU145 [21] and B16 melanoma [36] cells. In separate studies with nude mice, GA effectively inhibited growth of tumor in both DU145 and 22Rv1 prostate cancer xenografts, while it successfully decreased micro-vessel density, as compared to controls [37].

A series of in vitro experiments with HCT-15 intestinal carcinoma cells, demonstrated that growth was effectively inhibited by anthocyanin fractions extracted from different cherry and berry extracts in comparison to that of flavonoid fractions [38] while berry extracts including lingonberry, strawberry, blueberry, and bilberry extracts that contain anthocyanins inhibited the growth of HCT-116 colon cancer cells. Studies with Caco-2 colon cancer and HT-29 colonic crypt resembling cells demonstrated that flavonoids can significantly reduce proliferation. Dosing studies for the flavonoids showed anti-proliferative activity of all compounds with EC50 values ranging between 39.7 ± 2.3 microM (baicalein) and 203.6 ± 15.5 microM (diosmin) [39]. The anti-oxidant activity of the raspberry was directly related to the total amount of phenolics and flavonoids found in the raspberry ($p < 0.01$). No relationship was found between anti-proliferative activity and the total amount of phenolics/flavonoids found in the same raspberry ($p > 0.05$) suggesting that anti-proliferative capacity is maxed-out at a certain level of phenolics/flavonoids exerting their function, possibly through molecular signaling [40].

In terms of the form of berries, it has been demonstrated that wild grown species generally contain more phenolics than cultivated ones [41]. Measurement of the antioxidant activity of anthocyanin extracts in the case of blueberries for example, showed that there was no significant difference between fresh, dried, and frozen blueberries suggesting that the major compounds of interest are relatively stable and unaffected by general food industry typical processes and relevant production methods [42].

5. The Role of Dietary Berries in Various Types of Cancer

5.1. Cancers of the GI Tract

5.1.1. Diet and Colon Cancer—Berries/Phenolics and Colon Cancer

A close link is observed between colon carcinogenesis and chronic inflammation of the intestine in mouse models of inflammation and cancer [43–46].

Encouraging results from human studies pave the road for further clinical studies with black raspberry/berries especially considering the possibility of formulating berry-derived foods with consistent bioactive composition that can also be scaled-up for large clinical trials [47]. Berry phenolics, despite extensive metabolism and structural changes, seem to maintain their protective effects in relation to colon carcinogenesis [48].

In Vivo Studies

Protective effects of the mulberry fruit, containing flavonoids and anthocyanins, alkaloids and carotenoids, were documented by in vitro and in vivo studies on colon cancer and intestinal inflammation [49]. In MUC2($-/-$) mice, a model of spontaneous chronic intestinal inflammation at an early age and intestinal tumors at three months, a diet enriched with 5% or 10% mulberry extract administered for three months, starting at 3–4 weeks of age, elicited a reduction in tumorigenesis and intestinal inflammation as estimated by the degree of mucosal damage and lymphocyte infiltration [49]. In the same study, 6–8 week old BALB/c mice supplemented with mulberry extract (5% or 10% wt/wt) for 10 days before exposure to 3% dextran sulfate sodium (DSS) in drinking water for 9 days, showed an improvement in typical symptoms of DSS-induced acute colitis such as weight loss, bloody stools and colorectal histological changes [49]. In lipopolysaccharide (LPS)-treated RAW264.7 macrophages, mulberry extracts attenuated inflammation by reducing the expression of iNOS, COX-2, IL-1 beta and IL-6, and inhibiting activation of NF-κB/p65 and pERK/MAPK pathways [49]. Strawberry-enriched diets for 13 weeks were documented to inhibit chemically induced colorectal cancer in Crj: CD-1 mice after treatment with azoxymethane (AOM) and dextran sulfate sodium (DSS). Tumor incidence of the mice fed with freeze-dried strawberry powder at 2.5%, 5% and 10% (wt/wt) of the diet was measured at 64%, 75% and 44%, respectively, versus the 100% of their littermates fed the control diet. Tumor multiplicity was also reduced in all strawberry-fed mice, but reached statistically significant difference only with the 10% strawberry diet. Colon inflammation was also reduced due to the strawberry diet as observed by decreased nitrotyrosine, phosphorylation of PI3-kinase, Akt, ERK and NF-κB, expression of TNF-α, IL-1β, IL-6, iNOS and COX-2, as well as activity of iNOS and COX-2 [46].

Both animal and human studies have documented the anti-carcinogenic properties of bilberry anthocyanins [50,51]. In ApcMin mice, model of human familial adenomatous polyposis, a condition characterized by multiple intestinal polyps potentially developing to carcinomas, an anthocyanin-rich bilberry extract (40% anthocyanins) fed for 12 weeks at a dietary dose of 0.3% elicited a 30% reduction o adenoma counts in a dose-dependent manner [50].

In a pilot study, a standardized bilberry extract (36% anthocyanins wt/wt) was administered daily to 25 colorectal cancer patients for 7 days before scheduled resection of primary tumor or liver metastases. All the three different doses provided—1.4, 2.8, or 5.6 g (0.5–2.0 g anthocyanins)—were safe and well tolerated. Compared to pre-intervention, a significant 7% decrease of the proliferation index was observed in all colorectal tumors from all patients. In the lowest dose to patients, a significant 9% decrease in tumor tissue proliferation was measured, while with the other doses the observed decrease was not significant. In addition, serum insulin-like growth factor-1 levels were lower than pre-intervention in all patients, although not significantly [51]. Similarly, animal dietary studies with black raspberries reveal the chemopreventive capacities of the fruit in models of ulcerative colitis, several rodent models of colon cancer and adenoma, as well as humans [52–58].

Interleukin-10 knockout mice, model of ulcerative colitis-associated colon cancer progression, when fed with 5% wt/wt black raspberry presented diminished colon ulceration, along with lower levels of β-catenin nuclear translocation and inhibition of epigenetic events related to abnormal Wnt signaling [53]. Similarly, in DSS-induced ulcerative colitis in C57BL/6J mice, a 5% wt/wt black raspberry diet suppressed colon ulceration by restoring epigenetic regulation of systemic inflammation [59]. Intestinal tumor formation and cell proliferation was inhibited after a 12-week consumption of a Western-style diet supplemented with 10% (wt/wt) freeze-dried black raspberries in Apc1638+/− and Muc2−/− mice, both models of human colorectal cancer, although via distinct mechanisms. Tumor incidence and multiplicity was reduced by 45% and 60%, respectively in the Apc1638+/− mouse, and by 50% in the Muc2−/− mouse. A slight, non-significant reduction of tumor size was also observed. Signaling mediated by β-catenin and chronic inflammation, leading to colon pathology in the Apc and Muc2−/− mouse, respectively, were both attenuated due to the black raspberry diet. In the Apc1638+/− mouse, β-catenin was markedly reduced, while its downstream effectors, c-Myc and cyclin D1 were slightly reduced, along a positive, but non-significant modulation of several inflammatory markers (COX-2, TNF-α, IL-6, IL-10 and IL-1) [52].

A 5% wt/wt black raspberry supplementation for 8 weeks reduced polyp number and size in the Apc(Min/+) mice intestine and colon, inhibiting the development of colonic adenoma. Non-targeted metabolomics revealed that several apc-related metabolites in the mucosa, liver and feces were positively modulated by the black raspberry diet. Among the affected metabolites, putrescine and linolenate, are associated with colorectal cancer in humans [53]. In Fischer 344 rats consuming black raspberries at 2.5%, 5%, or 10% (wt/wt) of diet after AOM injections, aberrant crypt foci (ACF), total tumor and adenocarcinoma multiplicity were significantly reduced at varying percentages depending on the berry dose, while tumor burden showed a non-significant decrease across all berry diet groups. With regards to the dietary dose of black raspberries, the 5% dose was more effective than the 2.5% dose in inhibiting chemically induced adenocarcinoma, while the 10% dose did not produce any greater benefit. Finally, oxidative stress, as measured by urinary levels of 8-hydroxy-2'-deoxyguanosine (8-OHdG), was attenuated significantly in all rats supplemented with black raspberries [54].

The chemoprevention effect of three types of berries (bilberry, lingonberry and cloudberry) on intestinal tumorigenesis was tested in Min-mice by Misikangas et al. A diet containing 10% (wt/wt) freeze-dried bilberry, lingonberry or cloudberry fed for 10 weeks to mice produced a 10%–30% (statistically significant) reduction of intestinal carcinomas as well as reduced tumor burden by more than 60% [59]. As assessed by microarray analyses, with berry enriched diets the authors observed attenuation of genes implicated in colon carcinogenesis, including the decreased expression of the adenosine deaminase, ecto-5'-nucleotidase, and prostaglandin E2 receptor subtype EP4, thus suggesting some mechanistic evidence as to a potential explanation for the tumor risk reduction. In separate in vivo experiments also on Min mice, Rajakangas et al., showed that a 10% white currant dietary supplementation administered for 10 weeks can yield a significant reduction in the number and size of adenomas in the total small intestine, associated with reduced nuclear beta-catenin and NF-κB protein levels in the adenomas [60].

Many flavonoids have been shown to exert anti-carcinogenic effects in cells and animals. In many of these cases the compounds are administered in isolation or as part of a whole food yet processed, usually freeze-dried. Many of the chemopreventive effects observed seem to occur through signaling pathways known to be important in the pathogenesis of colorectal, gastric and esophageal cancers. Nonetheless, dietary flavonoid intakes are generally low and their metabolism in humans is extremely complex [61]. Additionally, the amounts and the optimum mixture of these protective compounds are unclear. Certain antioxidants exert undesirable pro-oxidant action when the dosing is high. As far as the initiation and progression of cancers of the gastrointestinal (GI) tract in humans, it is more probable that any adverse effects of diet are caused primarily by over-consumption of energy, coupled with inadequate intakes of protective substances, including micronutrients, dietary fiber and various phytochemicals. Carcinomas of the esophagus, stomach and colon all are suggested based on epidemiological observations to be partially preventable by diets rich in fruits and vegetables [62,63]. Hence, the consumption of berries is advisable in the context of risk reduction for colon cancer.

Human Studies

There is a limited number of human studies investigating the effect of edible berries on colon cancer, nevertheless the available clinical results are encouraging overall, indicating potential health benefits not only in terms of chemoprevention but also in regards to cancer patients' responses. More specifically, black raspberry was documented to reduce significantly cancer cell proliferation in 20 cancer patients, 6 with colon cancer and 14 with rectal cancer; 17 male and 3 female [56]. The study participants consumed three times daily 20 g of freeze-dried berry powder mixed with 100 mL water for varying periods of 1 to 9 weeks (4 weeks on average) [56]. It is estimated that daily consumption of 60 g powder is equivalent to 0.59 kg of fresh black raspberry and a rodent diet of approximately 7% wt/wt powder, a dose adequate for chemoprevention in animal studies [63]. The black raspberry treatment modified positively genetic and epigenetic markers measured in colorectal adenocarcinomas and adjacent normal tissues, as observed by modified gene expression of β-catenin and E-cadherin downstream of the Wnt pathway, and demethylation of *SFRP2* and *WIF*, tumor suppressor genes upstream of the Wnt pathway [56]. Aberrant signaling of Wnt/β-catenin pathway occurs in about 85% of sporadic colorectal cancers and is mainly attributed to Apc gene mutations [64]. In addition to Wnt pathway, black raspberry modified protectively expression of genes related to proliferation, apoptosis, and angiogenesis [56]. In 24 cancer patients receiving the same black raspberry treatment, plasma concentration of granulocyte macrophage colony stimulating factor (GM-CSF) was increased compared to measurements before initiating the berry treatment, while the other 8 plasma cytokines measured (IL-1β, IL-2, IL-6, IL-8, IL-10, IL-12p70, Interferon-γ, TNF-α) were not affected significantly. However, plasma concentrations of IL-8 decreased when the berry drink was consumed for longer than 10 days, suggesting that longer interventions may be needed for greater changes to occur [57]. In the same cohort of cancer patients non-targeted metabolomics in urine and plasma samples, showed significant changes in a number of metabolites (34 and 6, respectively) related to energy pathways, including increased carbohydrate and amino acid metabolites [57]. The black raspberry extract, which was reasonably well tolerated by cancer patients, overall showed promise as an ameliorating strategy in colon cancer cases.

Animal and human studies of two commercial products containing blackcurrant extract power, with or without lactoferrin and lutein, demonstrated protective effects of the berry supplement on the colonic microbiota, including changes in bacterial population and pH associated with colon cancer risk [65,66]. Sprague-Dawley rats, starting at 8 weeks of age, were fed the blackcurrant extract (13.4 mg/kg of body weight) three times weekly for a period of 4 weeks. A significant increase in beneficial populations of lactobacilli and bifidobacteria, along with increased activity of the related enzyme β-glucosidase, was attributed to the berry extract. In addition, the berry extract was shown to reduce bacteroides and clostridia, as well as the activity of β-glucuronidase, a bacterial enzyme involved in colon carcinogenesis [65]. Similarly, in thirty healthy volunteers (16 women, 14 men, age 20–60 years) the blackcurrant extract (672 mg/d for two weeks) elicited a positive modulation of gut microbiota by enhancing growth of lactobacilli and bifidobacteria (beneficial microflora), while lowering fecal pH, bacteroides and clostridia and inhibiting β-glucuronidase [66].

5.2. Esophageal Cancer

A significant amount of data, derived from in vivo work, demonstrate a potential set of benefits from berry consumption in terms of both chemoprevention and response to cancer in the case of esophageal cancer. Stoner and colleagues described the effects of black raspberries on gene expression in the very early stages of rat esophageal carcinogenesis and particularly showed effects on genes that influence carcinogen metabolism. More specifically, out of 2261 dysregulated genes in the esophagi of Fisher-344 rats treated for one week with nitrosomethylbenzylamine (NMBA), 462 were positively modulated and restored back to near-normal levels when a 5% (wt/wt) whole black raspberry (BRB) powder supplementation to the control AIN-76A diet was introduced [67]. The genes that were positively regulated were oncogenes, genes involved in oxidative damage and tumor suppressor genes regulating apoptosis, cell cycling and angiogenesis [67]. Furthermore, in other feeding experiments with F344 rats, animals were fed berry-supplemented diets for 2 weeks prior to NMBA treatments, and were then maintained on the berry-supplemented diets until the end of the 30-week experiment at which point esophageal tumors were counted. The rats were fed 5% (wt/wt) BRB powder, an anthocyanin-rich fraction, an organic solvent-soluble extract (each contained approximately 3.8 μmol anthocyanins/g diet), an organic-insoluble (residue) fraction (containing 0.02 μmol anthocyanins/g diet), a hexane extract, and a sugar fraction. The experiments indicated that anthocyanins derived from black raspberries were effective in reducing NMBA-induced tumors in the rat esophagus [68]. While several forms of anthocyanins were evaluated, all forms, i.e., 5% whole raspberries extract, anthocyanin-rich fraction and organic-solvent soluble extract, produced similar results. Interestingly, the organic insoluble residue fraction also produced comparable results, hence indicating that other compounds from berry anthocyanins may also extend a chemopreventive effect [68].

Alteration in the immune cell trafficking in esophageal cancer by anthocyanin constituents was shown in a rat feeding study [69]. After 5 weeks of NMBA (0.35 mg/kg) administration while rats were all on control diet, the animals were switched to the treatment diets (i.e., 6.1% black raspberries powder, anthocyanin-rich fraction of BRB 3.8 μmol/g, and 500 ppm protocatechuic acid). Esophageal cancer-related inflammatory biomarkers were assessed at three time-points (15, 25 and 35 weeks) in the plasma and at the esophagus. Furthermore, the infiltration of the immune cells in the esophagus was also evaluated. The production of cytokines was not different among all three different dietary treatments. However, all treatment groups exhibited a decrease of IL-1β and IL-12 and an increase of IL-10 compared to the control, while the treatments all produced a significantly lower infiltration of both macrophages and neutrophils to the esophagus. These data, taken together, suggest that anthocyanins inhibit esophageal tumorigenesis by altering cytokine expression and innate immune cell trafficking in the tumors [69].

The effect of black raspberries (whole food) in the late stages of rat esophageal carcinogenesis was investigated thoroughly by Wang and colleagues [70]. A 5% (wt/wt) freeze-dried raspberry-supplemented diet was evaluated on F344 rats versus control diet in terms of its influence in the late stages of carcinogenesis for NMBA-induced esophageal cancer. Rats were evaluated 35 weeks post-NMBA treatment and while on control versus raspberry-supplemented diets. Results showed that BRB reduced the number of dysplastic lesions as well as both the number and size of esophageal tumors. Furthermore, BRB was found to have a positive impact on the expression of a variety of genes associated with pre-neoplastic esophagus as well as esophageal papillomas, and found to modulate gene expression associated with proliferation, apoptosis and inflammation as well as angiogenesis in a positive manner. These results demonstrate that BRB can have a positive impact in terms of molecular signaling in the cases of even late stages carcinogenesis in esophageal cancer in rats, thus underlining potential benefits from black raspberry consumption in terms of cancer prevention and/or treatment.

Further evidence on the positive role of black raspberry consumption was provided by similar experiments conducted by Chen and colleagues [71]. A 5% (wt/wt) BRB diet was shown to reduce both mRNA and protein levels of COX-2, iNOS and c-Jun as well as level of prostaglandin E2 in F344 NMBA-treated rats on diets for 25 weeks. These findings suggest another potential role in cancer molecular signaling and possible therapeutic strategy in the case of esophageal cancer for black raspberries. The same group also reports a BRB angiogenesis-suppression parallel to the suppression of COX-2 and iNOS in separate experiments [72]. Other researchers observed similar positive effects of BRB in NMBA-induced esophageal cancer in F344 rats although the effects were not dose-dependent (5% and 10% wt/wt BRB supplementation produced similar effects) [73]. Such observations support the notion that finest dosing is optimizing health benefit, probably primarily through molecular signaling and hence more emphasis needs to be placed on the determination of the "ideal" mixture of bioactive compounds and dosing when it comes to dietary advice.

Stoner et al., [74] assessed the chemopreventive ability against esophageal cancer in rats, for a variety of berries. NMBA-treated rats over a period of 5 weeks were placed on diets supplemented with 5% (wt/wt) freeze-dried black or red raspberries, strawberries, blueberries, noni, açaí or wolfberry. The diets were provided for 35 weeks. The study revealed that all berry types were about equally effective in inhibiting NMBA-induced tumorigenesis in the rat esophagus although the content of anthocyanins and other comparable phytochemicals varies significantly among these berries. Interestingly, in all cases, the levels of serum cytokines, interleukin 5 (IL-5) and GRO/KC, the rat homologue for human interleukin-8 (IL-8) were notably reduced. These observations suggest that potential berry chemoprevention may be exerted via a variety of combinations in terms of type and amount of bioactive compounds thus not only underlining the importance of quality and synergism but also the importance of overall diet (as a pattern) in terms of diet-associated health benefits.

Kresty and colleagues demonstrated that cranberry proanthocyanidins inhibit esophageal adenocarcinoma both in vitro and in vivo. Specifically, purified cranberry-derived proanthocyanidin extract (C-PAC) was evaluated in its capacity to extend chemoprevention utilizing acid-sensitive and acid-resistant human esophageal adenocarcinoma (EAC) cell lines and esophageal tumor xenografts in athymic NU/NU mice [75]. Results showed pleiotropic cell death induction and PI3K/AKT/mTOR axis inactivation upon in vitro and in vivo exposure to C-PAC, hence indicating a potential mode of action for C-PAC chemoprevention.

5.3. Breast Cancer

Berry potential in reducing risk of breast cancer has been demonstrated by in vivo studies. More specifically, the preventive and therapeutic potential of highbush blueberry was investigated in (August-Copenhagen-Irish) ACI female rats supplemented with 5% wt/wt blueberry, 2 weeks before or 12 weeks after treatment with the carcinogen 17β-estradiol (E2), respectively. The tumor latency for palpable mammary tumors was delayed, while tumor volume and multiplicity was reduced in both intervention modes. A smaller dose of 2.5% blueberry diet administered after E2 treatment (i.e., therapeutically) in another experimental group also attenuated tumor multiplicity [76,77]. The aforementioned anti-tumor effects of blueberries were in agreement with favorable molecular changes such as down-regulation of CYP1A1 and ER-α gene expression, controlling E2 metabolism and signaling, respectively [76,77]. Blueberry and black raspberry were shown to possess protective effects against estrogen-induced breast cancer, although with varying impact and potential mechanism. At 5% wt/wt of diet, blueberry appeared more effective in reducing tissue proliferation, tumor burden and down-regulating CYP1A1 expression, while black raspberry delayed tumor latency to a greater extent and down-regulated ERα expression [77].

Notably, a series of phytochemicals abundantly found in berries such as cyanidin, delphinidin, quercetin, kaempferol, ellagic acid, resveratrol, and pterostilbene have been shown by in vitro and in vivo studies to interact and interfere with key pathways in breast cancer as well as induce apoptosis and autophagy thus reducing risk for breast cancer development and recurrence [78].

5.4. Miscellaneous Cancers and Berries

In addition to the most studied cancers (those of the GI tract and hormonal cancers), in terms of the role of berries in risk attenuation, impressive evidence is accumulating, suggesting that consumption of berries may well offer chemoprevention and/or improve responses in several types/cases of cancer. Several studies have demonstrated that strawberry consumption in rodents exerts chemoprevention in a series of cancers including oral cavity, breast, lung and esophageal [79–83]. In a phase II clinical trial strawberry consumption (60 g/day for 6 months) inhibited the progression of precancerous lesions [84]. Based on previous observations, potential mechanisms of anticancer activity include suppression of NF-κB, COX-2 and iNOS. Some evidence support a potential positive role of berries (specifically dietary polyphenols) in the case of prostate cancer [21] but there is no definitive conclusion as to the mechanisms of action, while results are sometimes inconsistent and variable.

Using the hamster cheek pouch model, Casto and colleagues investigated the potential of lyophilized strawberries to inhibit tumorigenesis. More specifically, animals were painted three times a week for six weeks with 0.2% 7,12-dimethylbenz(a)anthracene (DMBA) to induce oral cancer [79]. Hamsters were given diets supplemented with 5% or 10% wt/wt lyophilized strawberries prior to, during, and after, or only after carcinogen treatment. At 12 weeks after DMBA initiation, animals were terminated and number of lesions and tumors was counted [79]. Results revealed significantly fewer tumors and lesions in the strawberry-supplemented diet groups compared to controls, suggesting a potential positive effect of strawberries in the case of oral cancer. In separate experiments using the same animal model and carcinogen Zhu et al., demonstrated that tumor incidence, multiplicity, volume and histological grade of oral precancerous lesions were reduced in hamsters fed a 5% wt/wt lyophilized strawberry supplemented diet compared to animals fed control diet [84]. From a mechanistic perspective, the study showed that strawberries suppress cell proliferation, angiogenesis and oncogenic signaling and arachidonic acid metabolism [84].

Research conducted by Bishayee et al., showed that in rats fed a diet supplemented with pomegranate extract, PE (1 or 10 g/kg) 4 weeks before and 18 weeks following diethylnitrosamine (DENA)-initiated hepatocarcinogenesis, PE dose-dependently suppressed a series of elevated inflammatory markers (COX-2, NF-κB) [85].

Evidence to suggest that berry extracts can yield chemoprevention against lung cancer was provided by Balansky et al. Whole-body Swiss ICR mice were exposed to mainstream cigarette smoke, at birth and then daily for 4 months. Black chokeberry and strawberry aqueous extracts were given as the only source of drinking water, starting after weaning and continuing for 7 months. Both berry extracts inhibited lung adenomas' development [81].

Knobloch and co-workers in a phase 0 human study showed that black raspberries, when administered as freeze-dried powder supplement in oral torches to oral cancer biopsy-confirmed patients, improve gene expression profile associated with oral cancer [86]. Transcriptional biomarkers assessed showed significant improvement for the berry supplemented diet demonstrating marked reduction for the pro-inflammatory genes NFKB1, PTGS2 as well as pro-survival genes AURKA, BIRC5, EGFR all strongly associated with oral cancer risk and progression [86].

Work performed by Mallery et al., showed that in patients with oral intraepithelial neoplasia lesions, topical application of bioadhesive gels that contained 10% wt/wt freeze-dried black raspberries for 12 weeks resulted in statistically significant reduction in lesional sizes, histological grades, and loss of heterozygosity events compared to placebo controls [87].

In a data analysis of a case-control human study of 230 patients conducted in Italy, a favorable role of flavonoids and proanthocyanidins in gastric cancer was demonstrated by Rossi and colleagues [88]. In separate studies, Rossi et al., conducted a cohort study in Northern Italy between 1991 and 2008 with 326 cases of incident pancreatic cancer and respective controls, examining the relationship between flavonoids and pancreatic cancer risk. According to their results the analyses showed that dietary proanthocyanidins (mostly present in apples, pears and pulses), may convey some protection against pancreatic cancer risk [89].

The potential effect of diets on cancer risk has been studied widely. Even though there is no specific and defined diet that reduces cancer risk, diets rich in fruits and vegetables have been repeatedly found to be beneficial in terms of reducing cancer risk. In this context, the scientific consensus refers more to dietary habits or good dietary practices than a well-defined diet per se. However, there has been notable amount of evidence to suggest that adherence to a Mediterranean diet is associated with reduced risk of overall cancer mortality as well as reduced risk of incidence of several cancer types (especially cancers of the colorectum, GI tract, breast, stomach, pancreas, prostate, liver) according to observational studies [90,91]. Interestingly, the Mediterranean diet is a diet rich in fiber, fruits, vegetables and grains as well as olive oil and fish while from a compound standpoint rich in vitamins, antioxidants, flavonoids, anthocyanins, and other phytochemicals as well as ω-3 fatty acids. Furthermore, notably the Mediterranean diet is often viewed as a set of practices and a life-style much more so than a mere diet or dietary scheme. Interestingly, a study showed that if fruit and vegetable intake levels increased to 300 and 400 g/day respectively, that could lead to a significant reduction in gastric cancers, particularly in the developing countries [92].

Even though the mechanisms remain not completely clarified, there is convincing evidence from epidemiological and experimental studies that dietary factors are likely to have a major influence on the risk of several types of cancer, particularly of the GI tract [93–97]. In vivo experiments using mice showed that diet-induced obesity increases risk of colonic cancer while a signaling/modular implication of leptin (both a hormone and a cytokine) was also suggested by the data [96,97]. In a more global approach, berries have been shown to play a positive role in health protection in cases of chronic diseases other than cancer as well, where inflammation and oxidative stress are key initiators for the onset of a disease [98–100]. In this regard, placing more emphasis on the dietary constituents but also on the whole food and on dietary patterns that may confer health benefits and potential protection against chronic diseases is important.

6. Conclusions

Although the initiation, progression and development of cancer is a multi-factorial phenomenon, the contribution of diet to chemoprevention has been suggested by a bulk of evidence stemming from in vitro, in vivo and human studies. A particular interest has been placed on fruits and vegetables of which edible berries constitute an interesting sub-group, as they are rich in a variety of compounds that have been shown to exert favorable effects in several types of cancers. Anthocyanins, flavonoids, other antioxidants and a plethora of more or less well-defined phytochemicals with antioxidant and other properties seem to offer a notable arsenal that reduces risk of cancer as evidenced by in vivo and human work. The mechanistic details of the mode of action remain somewhat elusive although involvement of certain pathways has been demonstrated. These pathways are responsible for the modulation of different cellular processes, showing certain common signaling events such as arrest of cell cycle by increasing levels of cyclin-dependent kinase inhibitor proteins (CDIs) and inhibition of cyclins, induction of apoptosis via cytochrome c release, activation of caspases and down- or up-regulation of Bcl-2 family members, inhibition of survival/proliferation signals (Akt, MAPK, NF-κB) and inflammation (COX-2, TNF, IL secretion), as well as suppression of key proteins that are critically involved in angiogenesis and metastasis.

A significant amount of work to identify and delineate mechanistic details has been undertaken in cell lines, which although useful and indicative, must be corroborated by in vivo data. Thus in vitro data interpretation must be careful and cautiously extrapolated to the in vivo and much more so to the human level. The work we have summarized herein, primarily from in vivo and human studies, indicates favorable effects of berry consumption in a variety of cancers by mechanisms that involve oxidative stress, inflammation and the related signaling. It appears rather doubtful that a particular compound in berries is responsible for the health benefits these food items extend. It seems that whole food consumption (edible berries) may be optimizing the benefit arguably due to synergies. Furthermore, it is highly challenging to decipher and mimic the exact mixture of berry compounds, quality/types of compounds and amounts, or physicochemical conditions and matrix involvement in berries that confer the benefit.

Edible berries have been demonstrated to extend chemoprevention in cancer primarily of the GI tract as well as breast and to a lesser degree of liver, prostate, pancreas and lung. Notably, no negative effects have been reported by berry administration, thus making it a plausible and potentially useful dietary strategy to reduce risk of cancer and help cancer patients with disease prognosis.

Supplementary Materials: The following are available online at www.mdpi.com/2076-3921/5/4/37/s1. Table S1. Berries with selected nutrient and phytochemical profiles expressed in values per 100 g of edible portion as determined by USDA.

Conflicts of Interest: The authors have no conflict of interest to declare.

References

1. Stewart, B.W.; Wild, C.P. (Eds.) *World Cancer Report*; World Health Organization: Geneva, Switzerland; IARC Publication: Lyon, France, 2014.
2. Siegel, R.L.; Miller, K.D.; Ahmedin, J. Cancer Statistics 2016. *CA Cancer J. Clin.* **2016**, *66*, 7–30. [CrossRef] [PubMed]
3. Agency for Healthcare Research and Quality. Available online: www.ahrq.gov (accessed on 4 April 2016).
4. Center for Disease Control. Available online: www.cdc.gov (accessed on 4 April 2016).
5. Boivin, D.; Blanchette, M.; Barrette, S.; Moghrabi, A.; Béliveau, R. Inhibition of cancer cell proliferation and suppression of TNF-induced activation of NFkappaB by edible berry juice. *Anticancer Res.* **2007**, *27*, 937–948. [PubMed]
6. Block, G.; Patterson, B.; Subar, A. Fruit, vegetables, and cancer prevention: A review of the epidemiological evidence. *Nutr. Cancer* **1992**, *18*, 1–29. [CrossRef] [PubMed]
7. Steinmetz, K.A.; Potter, J.D. Vegetables, fruit, and cancer. I. Epidemiology. *Cancer Causes Control* **1991**, *2*, 325–357. [CrossRef] [PubMed]
8. Liu, R.H. Potential synergy of phytochemicals in cancer prevention: Mechanism of action. *J. Nutr.* **2004**, *134*, 3479S–3485S. [PubMed]
9. Chun, O.K.; Chung, S.J.; Claycombe, K.J.; Song, W.O. Serum C-reactive protein concentrations are inversely associated with dietary flavonoid intake in U.S. adults. *J. Nutr.* **2008**, *138*, 753–760. [PubMed]
10. Corley, J.; Kyle, J.A.; Starr, J.M.; McNeill, G.; Deary, I.J. Dietary factors and biomarkers of systemic inflammation in older people: The Lothian Birth Cohort 1936. *Br. J. Nutr.* **2015**, *114*, 1088–1098. [CrossRef] [PubMed]
11. Reuter, S.; Prasad, S.; Phromnoi, K.; Ravindran, J.; Sung, B.; Yadav, V.R.; Kannappan, R.; Chaturvedi, M.M.; Aggarwal, B.B. Thiocolchicoside exhibits anticancer effects through downregulation of NF-κB pathway and its regulated gene products linked to inflammation and cancer. *Cancer Prev. Res.* **2010**, *3*, 1462–1472. [CrossRef] [PubMed]
12. Reuter, S.; Gupta, S.C.; Chaturvedi, M.M.; Aggarwal, B.B. Oxidative stress, inflammation, and cancer: How are they linked? *Free Radic. Biol. Med.* **2010**, *49*, 1603–1616. [CrossRef] [PubMed]
13. Schetter, A.J.; Heegaard, N.H.; Harris, C.C. Inflammation and cancer: Interweaving microRNA, free radical, cytokine and p53 pathways. *Carcinogenesis* **2010**, *31*, 37–49. [CrossRef] [PubMed]
14. Coussens, L.M.; Werb, Z. Inflammation and cancer. *Nature* **2002**, *420*, 860–867. [CrossRef] [PubMed]
15. Aggarwal, B.B.; Vijayalekshmi, R.V.; Sung, B. Targeting inflammatory pathways for prevention and therapy of cancer: Short-term friend, long-term foe. *Clin. Cancer Res.* **2009**, *15*, 425–430. [CrossRef] [PubMed]
16. Lin, W.W.; Karin, M. A cytokine-mediated link between innate immunity, inflammation, and cancer. *J. Clin. Investig.* **2007**, *117*, 1175–1183. [CrossRef] [PubMed]
17. Yang, Z.J.; Chee, C.E.; Huang, S.; Sinicrope, F.A. The role of autophagy in cancer: Therapeutic implications. *Mol. Cancer Ther.* **2011**, *10*, 1533–1541. [CrossRef] [PubMed]
18. National Cancer Institute. Available online: www.cancer.gov (accessed on 4 April 2016).
19. Manson, M.M. Cancer prevention—The potential for diet to modulate molecular signalling. *Trends Mol. Med.* **2003**, *9*, 11–18. [CrossRef]
20. Hickey, M.; King, C. *The Cambridge Illustrated Glossary of Botanical Terms*, 1st ed.; Cambridge University Press: Cambridge, UK, 2001.
21. Lall, R.K.; Syed, D.N.; Adhami, V.M.; Khan, M.I.; Mukhtar, H. Dietary polyphenols in prevention and treatment of prostate cancer. *Int. J. Mol. Sci.* **2015**, *16*, 3350–3376. [CrossRef] [PubMed]
22. Aggarwal, B.B.; Shishodia, S. Molecular targets of dietary agents for prevention and therapy of cancer. *Biochem. Pharmacol.* **2006**, *71*, 1397–1421. [CrossRef] [PubMed]

23. Saw, C.L.; Guo, Y.; Yang, A.Y.; Paredes-Gonzalez, X.; Ramirez, C.; Pung, D.; Kong, A.N. The berry constituents quercetin, kaempferol, and pterostilbene synergistically attenuate reactive oxygen species: Involvement of the Nrf2-ARE signaling pathway. *Food Chem. Toxicol.* **2014**, *72*, 303–311. [CrossRef] [PubMed]

24. Afrin, S.; Giampieri, F.; Gasparrini, M.; Forbes-Hernandez, T.Y.; Varela-López, A.; Quiles, J.L.; Mezzetti, B.; Battino, M. Chemopreventive and Therapeutic Effects of Edible Berries: A Focus on Colon Cancer Prevention and Treatment. *Molecules* **2016**, *21*, 169. [CrossRef] [PubMed]

25. Ramos, S. Cancer chemoprevention and chemotherapy: Dietary polyphenols and signalling pathways. *Mol. Nutr. Food Res.* **2008**, *52*, 507–526. [CrossRef] [PubMed]

26. Wang, S.Y.; Jiao, H. Scavenging capacity of berry crops on superoxide radicals, hydrogen peroxide, hydroxyl radicals, and singlet oxygen. *J. Agric. Food Chem.* **2000**, *48*, 5677–5684. [CrossRef] [PubMed]

27. Lee, J.H.; Khor, T.O.; Shu, L.; Su, Z.Y.; Fuentes, F.; Kong, A.N. Dietary phytochemicals and cancer prevention: Nrf2 signaling, epigenetics, and cell death mechanisms in blocking cancer initiation and progression. *Pharmacol. Ther.* **2013**, *137*, 153–171. [CrossRef] [PubMed]

28. Li, L.; Wang, L.; Wu, Z.; Yao, L.; Wu, Y.; Huang, L.; Liu, K.; Zhou, X.; Gou, D. Anthocyanin-rich fractions from red raspberries attenuate inflammation in both RAW264.7 macrophages and a mouse model of colitis. *Sci. Rep.* **2014**, *4*, 6234. [CrossRef] [PubMed]

29. Chen, P.N.; Kuo, W.H.; Chiang, C.L.; Chiou, H.L.; Hsieh, Y.S.; Chu, S.C. Black rice anthocyanins inhibit cancer cells invasion via repressions of MMPs and u-PA expression. *Chem. Biol. Interact.* **2006**, *163*, 218–229. [CrossRef] [PubMed]

30. Reuter, S.; Gupta, S.C.; Park, B.; Goel, A.; Aggarwal, B.B. Epigenetic changes induced by curcumin and other natural compounds. *Genes Nutr.* **2011**, *6*, 93–108. [CrossRef] [PubMed]

31. Jaganathan, S.K.; Vellayappan, M.V.; Narasimhan, G.; Supriyanto, E.; Octorina Dewi, D.E.; Narayanan, A.L.; Balaji, A.; Subramanian, A.P.; Yusof, M. Chemopreventive effect of apple and berry fruits against colon cancer. *World J. Gastroenterol.* **2014**, *20*, 17029–17036. [CrossRef] [PubMed]

32. Veeriah, S.; Kautenburger, T.; Habermann, N.; Sauer, J.; Dietrich, H.; Will, F.; Pool-Zobel, B.L. Apple flavonoids inhibit growth of HT29 human colon cancer cells and modulate expression of genes involved in the biotransformation of xenobiotics. *Mol. Carcinog.* **2006**, *45*, 164–174. [CrossRef] [PubMed]

33. Kern, M.; Pahlke, G.; Balavenkatraman, K.K.; Böhmer, F.D.; Marko, D. Apple polyphenols affect protein kinase C activity and the onset of apoptosis in human colon carcinoma cells. *J. Agric. Food Chem.* **2007**, *55*, 4999–5006. [CrossRef] [PubMed]

34. Maurya, D.K.; Salvi, V.P.; Nair, C.K. Radiation protection of DNA by ferulic acid under in vitro and in vivo conditions. *Mol. Cell. Biochem.* **2005**, *280*, 209–217. [CrossRef] [PubMed]

35. Stoner, G.D.; Wang, L.S.; Casto, B.C. Laboratory and clinical studies of cancer chemoprevention by antioxidants in berries. *Carcinogenesis* **2008**, *29*, 1665–1674. [CrossRef] [PubMed]

36. Kim, Y.J. Antimelanogenic and antioxidant properties of gallic acid. *Biol. Pharm. Bull.* **2007**, *30*, 1052–1055. [CrossRef] [PubMed]

37. Kaur, M.; Velmurugan, B.; Rajamanickam, S.; Agarwal, R.; Agarwal, C. Gallic acid, an active constituent of grape seed extract, exhibits anti-proliferative, pro-apoptotic and anti-tumorigenic effects against prostate carcinoma xenograft growth in nude mice. *Pharm. Res.* **2009**, *26*, 2133–2140. [CrossRef] [PubMed]

38. Koide, T.; Kamei, H.; Hashimoto, Y.; Kojima, T.; Terabe, K.; Umeda, T. Influence of flavonoids on cell cycle phase as analyzed by flow-cytometry. *Cancer Biother. Radiopharm.* **1997**, *12*, 111–115. [CrossRef] [PubMed]

39. Kuntz, S.; Wenzel, U.; Daniel, H. Comparative analysis of the effects of flavonoids on proliferation, cytotoxicity, and apoptosis in human colon cancer cell lines. *Eur. J. Nutr.* **1999**, *38*, 133–142. [CrossRef] [PubMed]

40. Liu, M.; Li, X.Q.; Weber, C.; Lee, C.Y.; Brown, J.; Liu, R.H. Antioxidant and antiproliferative activities of raspberries. *J. Agric. Food Chem.* **2002**, *50*, 2926–2930. [CrossRef] [PubMed]

41. Mikulic-Petkovsek, M.; Schmitzer, V.; Slatnar, A.; Stampar, F.; Veberic, R. Composition of sugars, organic acids, and total phenolics in 25 wild or cultivated berry species. *J. Food Sci.* **2012**, *77*, 1064–1070. [CrossRef] [PubMed]

42. Lohachoompol, V.; Srzednicki, G.; Craske, J. The Change of Total Anthocyanins in Blueberries and Their Antioxidant Effect After Drying and Freezing. *J. Biomed. Biotechnol.* **2004**, *5*, 248–252. [CrossRef] [PubMed]

43. Westbrook, A.M.; Szakmary, A.; Schiestl, R.H. Mechanisms of intestinal inflammation and development of associated cancers: Lessons learned from mouse models. *Mutat. Res.* **2010**, *705*, 40–59. [CrossRef] [PubMed]

44. Terzić, J.; Grivennikov, S.; Karin, E.; Karin, M. Inflammation and colon cancer. *Gastroenterology* **2010**, *138*, 2101–2114. [CrossRef] [PubMed]
45. Shi, N.; Clinton, S.K.; Liu, Z.; Wang, Y.; Riedl, K.M.; Schwartz, S.J.; Zhang, X.; Pan, Z.; Chen, T. Strawberry phytochemicals inhibit azoxymethane/dextran sodium sulfate-induced colorectal carcinogenesis in Crj: CD-1 mice. *Nutrients* **2015**, *7*, 1696–1715. [CrossRef] [PubMed]
46. Gu, J.; Ahn-Jarvis, J.H.; Riedl, K.M.; Schwartz, S.J.; Clinton, S.K.; Vodovotz, Y. Characterization of black raspberry functional food products for cancer prevention human clinical trials. *J. Agric. Food Chem.* **2014**, *62*, 3997–4006. [CrossRef] [PubMed]
47. Brown, E.M.; McDougall, G.J.; Stewart, D.; Pereira-Caro, G.; González-Barrio, R.; Allsopp, P.; Magee, P.; Crozier, A.; Rowland, I.; Gill, C.I. Persistence of anticancer activity in berry extracts after simulated gastrointestinal digestion and colonic fermentation. *PLoS ONE* **2012**, *7*, e49740. [CrossRef] [PubMed]
48. Qian, Z.; Wu, Z.; Huang, L.; Qiu, H.; Wang, L.; Li, L.; Yao, L.; Kang, K.; Qu, J.; Wu, Y.; et al. Mulberry fruit prevents LPS-induced NF-κB/pERK/MAPK signals in macrophages and suppresses acute colitis and colorectal tumorigenesis in mice. *Sci. Rep.* **2015**, *5*, 17348. [CrossRef] [PubMed]
49. Cooke, D.; Schwarz, M.; Boocock, D.; Winterhalter, P.; Steward, W.P.; Gescher, A.J.; Marczylo, T.H. Effect of cyanidin-3-glucoside and an anthocyanin mixture from bilberry on adenoma development in the ApcMin mouse model of intestinal carcinogenesis-relationship with tissue anthocyanin levels. *Int. J. Cancer* **2006**, *119*, 2213–2220. [CrossRef] [PubMed]
50. Thomasset, S.; Berry, D.P.; Cai, H.; West, K.; Marczylo, T.H.; Marsden, D.; Brown, K.; Dennison, A.; Garcea, G.; Miller, A.; et al. Pilot study of oral anthocyanins for colorectal cancer chemoprevention. *Cancer Prev. Res.* **2009**, *2*, 625–633. [CrossRef] [PubMed]
51. Bi, X.; Fang, W.; Wang, L.S.; Stoner, G.D.; Yang, W. Black raspberries inhibit intestinal tumorigenesis in apc1638+/− and Muc2−/− mouse models of colorectal cancer. *Cancer Prev. Res.* **2010**, *3*, 1443–1450. [CrossRef] [PubMed]
52. Wang, L.S.; Kuo, C.T.; Huang, T.H.; Yearsley, M.; Oshima, K.; Stoner, G.D.; Yu, J.; Lechner, J.F.; Huang, Y.W. Black raspberries protectively regulate methylation of Wnt pathway genes in precancerous colon tissue. *Cancer Prev. Res.* **2013**, *6*, 1317–1327. [CrossRef] [PubMed]
53. Pan, P.; Skaer, C.W.; Wang, H.T.; Stirdivant, S.M.; Young, M.R.; Oshima, K.; Stoner, G.D.; Lechner, J.F.; Huang, Y.W.; Wang, L.S. Black raspberries suppress colonic adenoma development in ApcMin/+ mice: Relation to metabolite profiles. *Carcinogenesis* **2015**, *36*, 1245–1253. [CrossRef] [PubMed]
54. Harris, G.K.; Gupta, A.; Nines, R.G.; Kresty, L.A.; Habib, S.G.; Frankel, W.L.; La Perle, K.; Gallaher, D.D.; Schwartz, S.J.; Stoner, G.D. Effects of lyophilized black raspberries on azoxymethane-induced colon cancer and 8-hydroxy-2'-deoxyguanosine levels in the Fischer 344 rat. *Nutr. Cancer* **2001**, *40*, 125–133. [CrossRef] [PubMed]
55. Wang, L.S.; Arnold, M.; Huang, Y.W.; Sardo, C.; Seguin, C.; Martin, E.; Huang, T.H.; Riedl, K.; Schwartz, S.; Frankel, W.; et al. Modulation of genetic and epigenetic biomarkers of colorectal cancer in humans by black raspberries: A phase I pilot study. *Clin. Cancer Res.* **2011**, *17*, 598–610. [CrossRef] [PubMed]
56. Pan, P.; Skaer, C.W.; Stirdivant, S.M.; Young, M.R.; Stoner, G.D.; Lechner, J.F.; Huang, Y.W.; Wang, L.S. Beneficial Regulation of Metabolic Profiles by Black Raspberries in Human Colorectal Cancer Patients. *Cancer Prev. Res.* **2015**, *8*, 743–750. [CrossRef] [PubMed]
57. Mentor-Marcel, R.A.; Bobe, G.; Sardo, C.; Wang, L.S.; Kuo, C.T.; Stoner, G.; Colburn, N.H. Plasma cytokines as potential response indicators to dietary freeze-dried black raspberries in colorectal cancer patients. *Nutr. Cancer* **2012**, *64*, 820–825. [CrossRef] [PubMed]
58. Wang, L.S.; Kuo, C.T.; Stoner, K.; Yearsley, M.; Oshima, K.; Yu, J.; Huang, T.H.; Rosenberg, D.; Peiffer, D.; Stoner, G.; et al. Dietary black raspberries modulate DNA methylation in dextran sodium sulfate (DSS)-induced ulcerative colitis. *Carcinogenesis* **2013**, *34*, 2842–2850. [CrossRef] [PubMed]
59. Misikangas, M.; Pajari, A.M.; Päivärinta, E.; Oikarinen, S.I.; Rajakangas, J.; Marttinen, M.; Tanayama, H.; Törrönen, R.; Mutanen, M. Three Nordic berries inhibit intestinal tumorigenesis in multiple intestinal neoplasia/+ mice by modulating beta-catenin signaling in the tumor and transcription in the mucosa. *J. Nutr.* **2007**, *137*, 2285–2290. [PubMed]
60. Rajakangas, J.; Misikangas, M.; Päivärinta, E.; Mutanen, M. Chemoprevention by white currant is mediated by the reduction of nuclear beta-catenin and NF-kappaB levels in Min mice adenomas. *Eur. J. Nutr.* **2008**, *47*, 115–122. [CrossRef] [PubMed]

61. Pierini, R.; Gee, J.M.; Belshaw, N.J.; Johnson, I.T. Flavonoids and intestinal cancers. *Br. J. Nutr.* **2008**, *99*, ES53–ES59. [CrossRef] [PubMed]

62. Johnson, I.T. New approaches to the role of diet in the prevention of cancers of the alimentary tract. *Mutat. Res.* **2004**, *551*, 9–28. [CrossRef] [PubMed]

63. Stoner, G.D.; Wang, L.S.; Zikri, N.; Chen, T.; Hecht, S.S.; Huang, C.; Sardo, C.; Lechner, J.F. Cancer prevention with freeze-dried berries and berry components. *Semin. Cancer Biol.* **2007**, *17*, 403–410. [CrossRef] [PubMed]

64. Klaus, A.; Birchmeier, W. Wnt signalling and its impact on development and cancer. *Nat. Rev. Cancer* **2008**, *8*, 387–398. [CrossRef] [PubMed]

65. Molan, A.L.; Liu, Z.; Kruger, M. The ability of blackcurrant extracts to positively modulate key markers of gastrointestinal function in rats. *World J. Microbiol. Biotechnol.* **2010**, *26*, 1735–1743. [CrossRef]

66. Molan, A.L.; Liu, Z.; Plimmer, G. Evaluation of the effect of blackcurrant products on gut microbiota and on markers of risk for colon cancer in humans. *Phytother. Res.* **2014**, *28*, 416–422. [CrossRef] [PubMed]

67. Stoner, G.D.; Dombkowski, A.A.; Reen, R.K.; Cukovic, D.; Salagrama, S.; Wang, L.S.; Lechner, J.F. Carcinogen-altered genes in rat esophagus positively modulated to normal levels of expression by both phenethyl isothiocyanate and black raspberries. *Cancer Res.* **2008**, *68*, 6460–6467. [CrossRef] [PubMed]

68. Wang, L.S.; Hecht, S.S.; Carmella, S.G.; Yu, N.; Larue, B.; Henry, C.; McIntyre, C.; Rocha, C.; Lechner, J.F.; Stoner, G.D. Anthocyanins in black raspberries prevent esophageal tumors in rats. *Cancer Prev. Res.* **2009**, *2*, 84–93. [CrossRef] [PubMed]

69. Peiffer, D.S.; Wang, L.S.; Zimmerman, N.P.; Ransom, B.W.; Carmella, S.G.; Kuo, C.T.; Chen, J.H.; Oshima, K.; Huang, Y.W.; Hecht, S.S.; et al. Dietary Consumption of Black Raspberries or Their Anthocyanin Constituents Alters Innate Immune Cell Trafficking in Esophageal Cancer. *Cancer Immunol. Res.* **2016**, *4*, 72–82. [CrossRef] [PubMed]

70. Wang, L.S.; Dombkowski, A.A.; Seguin, C.; Rocha, C.; Cukovic, D.; Mukundan, A.; Henry, C.; Stoner, G.D. Mechanistic basis for the chemopreventive effects of black raspberries at a late stage of rat esophageal carcinogenesis. *Mol. Carcinog.* **2011**, *50*, 291–300. [CrossRef] [PubMed]

71. Chen, T.; Hwang, H.; Rose, M.E.; Nines, R.G.; Stoner, G.D. Chemopreventive properties of black raspberries in *N*-nitrosomethylbenzylamine-induced rat esophageal tumorigenesis: Down-regulation of cyclooxygenase-2, inducible nitric oxide synthase, and c-Jun. *Cancer Res.* **2006**, *66*, 2853–2859. [CrossRef] [PubMed]

72. Chen, T.; Rose, M.E.; Hwang, H.; Nines, R.G.; Stoner, G.D. Black raspberries inhibit *N*-nitrosomethylbenzylamine (NMBA)-induced angiogenesis in rat esophagus parallel to the suppression of COX-2 and iNOS. *Carcinogenesis* **2006**, *27*, 2301–2307. [CrossRef] [PubMed]

73. Reen, R.K.; Nines, R.; Stoner, G.D. Modulation of *N*-nitrosomethylbenzylamine metabolism by black raspberries in the esophagus and liver of Fischer 344 rats. *Nutr. Cancer* **2006**, *54*, 47–57. [CrossRef] [PubMed]

74. Stoner, G.D.; Wang, L.S.; Seguin, C.; Rocha, C.; Stoner, K.; Chiu, S.; Kinghorn, A.D. Multiple berry types prevent *N*-nitrosomethylbenzylamine-induced esophageal cancer in rats. *Pharm. Res.* **2010**, *27*, 1138–1145. [CrossRef] [PubMed]

75. Kresty, L.A.; Weh, K.M.; Zeyzus-Johns, B.; Perez, L.N.; Howell, A.B. Cranberry proanthocyanidins inhibit esophageal adenocarcinoma in vitro and in vivo through pleiotropic cell death induction and PI3K/AKT/mTOR inactivation. *Oncotarget* **2015**, *6*, 33438–33455. [PubMed]

76. Jeyabalan, J.; Aqil, F.; Munagala, R.; Annamalai, L.; Vadhanam, M.V.; Gupta, R.C. Chemopreventive and therapeutic activity of dietary blueberry against estrogen-mediated breast cancer. *J. Agric. Food Chem.* **2014**, *62*, 3963–3971. [CrossRef] [PubMed]

77. Ravoori, S.; Vadhanam, M.V.; Aqil, F.; Gupta, R.C. Inhibition of estrogen-mediated mammary tumorigenesis by blueberry and black raspberry. *J. Agric. Food Chem.* **2012**, *60*, 5547–5555. [CrossRef] [PubMed]

78. Aiyer, H.S.; Warri, A.M.; Woode, D.R.; Hilakivi-Clarke, L.; Clarke, R. Influence of berry polyphenols on receptor signaling and cell-death pathways: Implications for breast cancer prevention. *J. Agric. Food Chem.* **2012**, *60*, 5693–5708. [CrossRef] [PubMed]

79. Casto, B.C.; Knobloch, T.J.; Galioto, R.L.; Yu, Z.; Accurso, B.T.; Warner, B.M. Chemoprevention of oral cancer by lyophilized strawberries. *Anticancer Res.* **2013**, *33*, 4757–4566. [PubMed]

80. Somasagara, R.R.; Hegde, M.; Chiruvella, K.K.; Musini, A.; Choudhary, B.; Raghavan, S.C. Extracts of strawberry fruits induce intrinsic pathway of apoptosis in breast cancer cells and inhibits tumor progression in mice. *PLoS ONE* **2012**, *7*, e47021. [CrossRef] [PubMed]

81. Balansky, R.; Ganchev, G.; Iltcheva, M.; Kratchanova, M.; Denev, P.; Kratchanov, C.; Polasa, K.; D'Agostini, F.; Steele, V.E.; de Flora, S. Inhibition of lung tumor development by berry extracts in mice exposed to cigarette smoke. *Int. J. Cancer* **2012**, *131*, 1991–1997. [CrossRef] [PubMed]

82. Carlton, P.S.; Kresty, L.A.; Siglin, J.C.; Morse, M.A.; Lu, J.; Morgan, C.; Stoner, G.D. Inhibition of *N*-nitrosomethylbenzylamine-induced tumorigenesis in the rat esophagus by dietary freeze-dried strawberries. *Carcinogenesis* **2001**, *22*, 441–446. [CrossRef] [PubMed]

83. Chen, T.; Yan, F.; Qian, J.; Guo, M.; Zhang, H.; Tang, X.; Chen, F.; Stoner, G.D.; Wang, X. Randomized phase II trial of lyophilized strawberries in patients with dysplastic precancerous lesions of the esophagus. *Cancer Prev. Res.* **2012**, *5*, 41–50. [CrossRef] [PubMed]

84. Zhu, X.; Xiong, L.; Zhang, X.; Chen, T. Lyophilized strawberries prevent 7,12-dimethylbenz[α]anthracene (DMBA)-induced oral squamous cell carcinogenesis in hamsters. *J. Funct. Foods* **2015**, *15*, 476–486. [CrossRef]

85. Bishayee, A.; Thoppil, R.J.; Darvesh, A.S.; Ohanyan, V.; Meszaros, J.G.; Bhatia, D. Pomegranate phytoconstituents blunt the inflammatory cascade in a chemically induced rodent model of hepatocellular carcinogenesis. *J. Nutr. Biochem.* **2013**, *24*, 178–187. [CrossRef] [PubMed]

86. Knobloch, T.J.; Uhrig, L.K.; Pearl, D.K.; Casto, B.C.; Warner, B.M.; Clinton, S.K.; Sardo-Molmenti, C.L.; Ferguson, J.M.; Daly, B.T.; Riedl, K.; et al. Suppression of Proinflammatory and Prosurvival Biomarkers in Oral Cancer Patients Consuming a Black Raspberry Phytochemical-Rich Troche. *Cancer Prev. Res.* **2016**, *9*, 159–171. [CrossRef] [PubMed]

87. Mallery, S.R.; Tong, M.; Shumway, B.S.; Curran, A.E.; Larsen, P.E.; Ness, G.M.; Kennedy, K.S.; Blakey, G.H.; Kushner, G.M.; Vickers, A.M.; et al. Topical application of a mucoadhesive freeze-dried black raspberry gel induces clinical and histologic regression and reduces loss of heterozygosity events in premalignant oral intraepithelial lesions: Results from a multicentered, placebo-controlled clinical trial. *Clin. Cancer Res.* **2014**, *20*, 1910–1924. [PubMed]

88. Rossi, M.; Rosato, V.; Bosetti, C.; Lagiou, P.; Parpinel, M.; Bertuccio, P.; Negri, E.; La Vecchia, C. Flavonoids, proanthocyanidins, and the risk of stomach cancer. *Cancer Causes Control* **2010**, *21*, 1597–1604. [CrossRef] [PubMed]

89. Rossi, M.; Lugo, A.; Lagiou, P.; Zucchetto, A.; Polesel, J.; Serraino, D.; Negri, E.; Trichopoulos, D.; La Vecchia, C. Proanthocyanidins and other flavonoids in relation to pancreatic cancer: A case-control study in Italy. *Ann. Oncol.* **2012**, *23*, 1488–1493. [CrossRef] [PubMed]

90. Schwingshackl, L.; Hoffmann, G. Does a Mediterranean-Type Diet Reduce Cancer Risk? *Curr. Nutr. Rep.* **2016**, *5*, 9–17. [CrossRef] [PubMed]

91. Grosso, G.; Buscemi, S.; Galvano, F.; Mistretta, A.; Marventano, S.; La Vela, V.; Drago, F.; Gangi, S.; Basile, F.; Biondi, A. Mediterranean diet and cancer: Epidemiological evidence and mechanism of selected aspects. *BMC Surg.* **2013**, *13* (Suppl. 2), S14. [CrossRef] [PubMed]

92. Peleteiro, B.; Padrão, P.; Castro, C.; Ferro, A.; Morais, S.; Lunet, N. Worldwide burden of gastric cancer in 2012 that could have been prevented by increasing fruit and vegetable intake and predictions for 2025. *Br. J. Nutr.* **2016**, *115*, 851–859. [CrossRef] [PubMed]

93. Sikalidis, A.K. Amino Acids and Immune Response: A role for cysteine, glutamine, phenylalanine, tryptophan and arginine in T-cell function and cancer? *Pathol. Oncol. Res.* **2015**, *21*, 9–17. [CrossRef] [PubMed]

94. Sikalidis, A.K.; Varamini, B. Roles of hormones and signaling molecules in describing the relationship between obesity and colon cancer. *Pathol. Oncol. Res.* **2011**, *17*, 785–790. [CrossRef] [PubMed]

95. Franco, A.; Sikalidis, A.K.; Solis Herruzo, J.A. Colorectal cancer: influence of diet and lifestyle factors. *Rev. Esp. Enferm. Dig.* **2005**, *97*, 432–448. [CrossRef] [PubMed]

96. Sikalidis, A.K.; Fitch, M.D.; Fleming, S.E. Risk of Colonic Cancer is Not Higher in the Obese Lep^{ob} Mouse Model Compared to Lean Littermates. *Pathol. Oncol. Res.* **2013**, *19*, 867–874. [CrossRef] [PubMed]

97. Sikalidis, A.K.; Fitch, M.D.; Fleming, S.E. Diet Induced Obesity Increases the Risk of Colonic Tumorigenesis in Mice. *Pathol. Oncol. Res.* **2013**, *19*, 657–666. [CrossRef] [PubMed]

98. Vendrame, S.; Daugherty, A.; Kristo, A.S.; Riso, P.; Klimis-Zacas, D. Wild blueberry (*Vaccinium angustifolium*) consumption improves inflammatory status in the obese Zucker rat model of the metabolic syndrome. *J. Nutr. Biochem.* **2013**, *24*, 1508–1512. [CrossRef] [PubMed]

99. Kristo, A.S.; Matthan, N.R.; Lichtenstein, A.H. Effect of diets differing in glycemic index and glycemic load on cardiovascular risk factors: Review of randomized controlled-feeding trials. *Nutrients* **2013**, *5*, 1071–1080. [CrossRef] [PubMed]
100. Vendrame, S.; Kristo, A.S.; Schuschke, D.A.; Klimis-Zacas, D. Wild blueberry consumption affects arterial vascular function in the obese Zucker rat. *Appl. Physiol. Nutr. Metab.* **2014**, *39*, 255–261. [CrossRef] [PubMed]

antioxidants

MDPI

Review

Cranberries and Cancer: An Update of Preclinical Studies Evaluating the Cancer Inhibitory Potential of Cranberry and Cranberry Derived Constituents

Katherine M. Weh [1], Jennifer Clarke [2,3,4] and Laura A. Kresty [1,*]

[1] Department of Medicine, Division of Hematology and Oncology, Medical College of Wisconsin, 8701 Watertown Plank Road, Milwaukee, WI 53226, USA; kweh@mcw.edu
[2] Department of Food Science and Technology, University of Nebraska, 256 Food Innovation Complex, Lincoln, NE 68588-6205, USA; jclarke3@unl.edu
[3] Department of Statistics, University of Nebraska, Lincoln, NE 68583, USA
[4] Quantitative Life Sciences Initiative, University of Nebraska, Lincoln, NE 68583, USA
* Correspondance: lkresty@mcw.edu; Tel.: +1-414-955-2673

Academic Editor: Dorothy Klimis-Zacas
Received: 30 June 2016; Accepted: 15 August 2016; Published: 18 August 2016

Abstract: Cranberries are rich in bioactive constituents reported to influence a variety of health benefits, ranging from improved immune function and decreased infections to reduced cardiovascular disease and more recently cancer inhibition. A review of cranberry research targeting cancer revealed positive effects of cranberries or cranberry derived constituents against 17 different cancers utilizing a variety of in vitro techniques, whereas in vivo studies supported the inhibitory action of cranberries toward cancers of the esophagus, stomach, colon, bladder, prostate, glioblastoma and lymphoma. Mechanisms of cranberry-linked cancer inhibition include cellular death induction via apoptosis, necrosis and autophagy; reduction of cellular proliferation; alterations in reactive oxygen species; and modification of cytokine and signal transduction pathways. Given the emerging positive preclinical effects of cranberries, future clinical directions targeting cancer or premalignancy in high risk cohorts should be considered.

Keywords: cranberry; cancer; proanthocyanidin; quercetin; ursolic acid

1. Introduction

Incorporation of fruit and vegetables, including cranberries, into a healthy-balanced diet is suggested for prevention of human disease. The positive health benefits of cranberries and cranberry derived constituents include improvements of cardiovascular function as measured by decreases in lipid peroxidation, oxidative stress, total and low-density lipoprotein (LDL) cholesterol and high-density lipoprotein (HDL) cholesterol level increases [1,2]. Cranberry derived products can also increase immune function by increasing $\gamma\delta$-T cells, NK cells and B-cells [3], as well as exhibit antimicrobial and anti-adhesion activities against Gram-positive bacteria [4], Gram-negative bacteria [5–11] and yeast [12,13]. Utilization of cranberry and cranberry derived constituents in the prevention of cancer is an underexplored area, but one with mounting preclinical in vitro and in vivo research as will be reviewed herein. To date, there have been no clinical trials conducted which utilize cranberries to prevent or delay cancer progression.

The beneficial effects of cranberries are attributable to the berries' rich phytonutrient composition which has been extensively and expertly reviewed by Pappas et al. [14]. Compositional analysis of the cranberry has resulted in identification and characterization of over 150 different bioactive constituents and human metabolomic studies have revealed differential pharmacokinetic profiles for these molecules [14–17]. Included in the family of polyphenols are three flavonoid classes:

anthocyanins, flavonols and proanthocyanidins. Specifically, flavonoids and phenolic acids are detected in the urine and plasma of healthy older adults following a single dose of 54% cranberry juice [15]. Cranberries and cranberry derived constituents are capable of exerting antioxidant and anti-inflammatory functions as supported by several clinical trials investigating cardiovascular health improvements measured via increases in flow mediated dilation, total antioxidant performance of plasma, blood glutathione peroxidase levels and superoxide dismutase activity following consumption of a cranberry juice cocktail [15,18–21].

The cancer inhibitory potential of cranberries and cranberry derived products is being elucidated based on multiple in vitro investigations and a small number of in vivo studies. This review will encompass a total of 34 preclinical studies utilizing 45 cancer cell lines isolated from 16 target organs and studies targeting seven cancers utilizing in vivo carcinogenesis and xenograft models to investigate mechanisms by which cranberries and cranberry derived constituents modulate or inhibit cancer-related processes. Mechanisms of cranberry-linked cancer inhibition are summarized in Figure 1. Preclinical studies support that cranberries modulate cell viability, cell proliferation, cell death, adhesion, inflammation, oxidative stress and signal transduction pathways. Many of the in vitro studies initially focus on the effectiveness of cranberry derived constituents in cell density and viability assays, as logical starting points for determining whether further mechanistic analysis is warranted. Collectively, these in vitro studies provide the fundamental basis for additional in vivo studies and may inform the design and implementation of cancer-based clinical trials evaluating cranberries as cancer preventive agents.

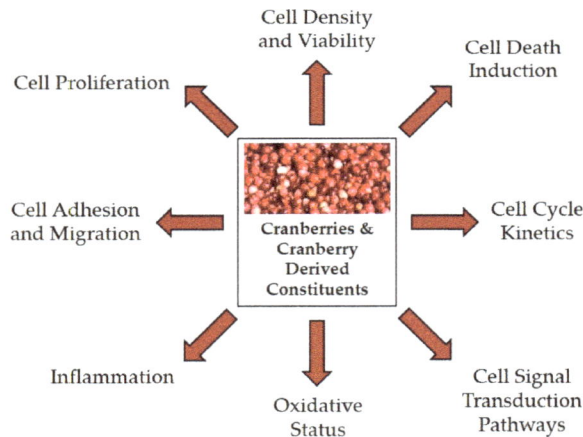

Figure 1. Cranberry and cranberry derived constituents target numerous mechanisms of cancer inhibition based on 34 preclinical studies.

2. Materials and Methods

A thorough bibliographic search was conducted in Pubmed through 4 June 2016 to identify all cancer focused research utilizing cranberries or cranberry derivatives. Keyword searches were performed by searching cranberry and each individual cancer target: breast, cervical, colon, esophageal, glioblastoma, leukemia, liver cancer, lung, lymphoma, melanoma, neuroblastoma, oral cavity, ovarian, prostate, renal/kidney, stomach and bladder. A secondary search was conducted using the same keywords in Scopus, an abstract and citation database for peer-reviewed literature, which yielded additional manuscripts not available in Pubmed. Finally, a bibliographic search was completed in the Health Research Library provided by the Cranberry Institute using the following keywords: cancer, reactive oxygen species, anti-oxidant and oxidative stress.

3. In Vitro Inhibition of Cancer Processes by Cranberries

Utilization of cancer cell lines to test cranberries, cranberry derived constituents and extracts has been the initial basis for defining cancer inhibitory capacity of these natural components in vitro. The majority of studies have been performed following treatment of immortalized cancer cell lines with cranberry extracts or juices, while one study has determined the benefit of pretreating the cells first to measure protective capacity against oxidative stress [22]. As summarized in Table 1, there are 31 in vitro based published reports describing cranberry linked cancer inhibition in 45 cancer cell lines derived from 16 targets. Eight mechanisms will be discussed with respect to cranberry derived extracts and constituents including cell density and viability, cell proliferation, cell cycle kinetics, cell death, signaling pathways, adhesion and migration, oxidative status and inflammation.

3.1. Cranberry Derived Extracts and Constituents Affect Cellular Growth and Viability

A significant amount of research shows cranberry derived constituents decrease cancer cell density, viability and proliferation. Cell density experiments are primarily based on treatment of cell lines with cranberry derived extracts followed by crystal violet staining to visualize cells remaining after treatment. While crystal violet staining indicates a qualitative difference in cellular confluency based on treatment, it does not rely on active metabolic processes. Viability stains including 3-(4,5-dimethylthiazol-2-yl)-2,5-diphenyltetrazolium bromide (MTT) and 2-(4-iodophenyl)-3-(4-nitrophenyl)-5-(2,4-disulfophenyl)-2H-tetrazolium, monosodium salt (WST-1) rely on cleavage of tetrazolium salts by the succinate-tetrazolium reductase system in metabolically active cells with an intact mitochondrial respiratory chain, whereas Calcein-acetoxymethyl (Calcein-AM) is cleaved by intracellular esterases to a fluorescent molecule, which results in more reliable data when cell death involves altered mitochondrial machinery. Bromodeoxyuridine (BrdU) incorporation assays as a measurement of cell proliferation inform how cranberry derived constituents modulate S-phase cell cycle kinetics.

Table 1. Summary of preclinical in vitro evaluations of cranberries or cranberry derived constituents as cancer inhibitors.

Target	Cell Line(s)	Cranberry Constituent	In Vitro Results [Reference(s)]
Breast	MCF-7	CE	↑ apoptosis [23]; ↑ G_1 cell cycle arrest [23] ↓ cell viability [23,24]
		CJE	↓ cell viability [25]
		C-PAC	↓ cell density [26]
		FG	↓ cell viability [27]
		Fr6	↓ cell viability [28]
		Q	↓ cell viability [27]
		UA	↓ cell density [29,30]; ↓ cell viability [27]
	MDA-MB-435*	CJE	↓ cell viability [25]
		Fr6	↑ apoptosis [28]; ↑ G_2-M cell cycle arrest [28] ↓ cell viability [28]
		UA	↓ cell density [29]
Cervix	ME180	C-PAC	↓ cell density [26]
		UA	↓ cell density [30]

Table 1. *Cont.*

Target	Cell Line(s)	Cranberry Constituent	In Vitro Results [Reference(s)]
Colon	Caco-2	CJE	↓ cell viability [25]
		TP	↓ lipid peroxidation [31]
			↓ pro-inflammatory markers TNFα and IL-6 [31]
	HT-29	ANTHO	↓ cell viability [32]
		CE	↓ cell viability [24,33]
			↓ pro-inflammatory marker COX-2 [34]
		C-PAC	↑ apoptosis [35]; ↓ cell density [26]
			↓ cell viability [36]
		CJE	↓ cell viability [32]
		Fr6	↓ cell viability [36]
		TP	↓ cell viability [32]
		UA	↑ apoptosis [35]; ↓ cell density [29,30]
			↓ cell viability [35]
	HCT116	CE	↓ cell viability [33]
		C-PAC	↑ apoptosis [35]; ↓ cell viability [35]
		UA	↑ apoptosis [35]; ↓ cell density [29]
			↓ cell viability [35]
	LS-513	ANTHO	↓ cell viability [32]
		CJE	↓ cell viability [32]
		TP	↓ cell viability [32]
	SW460	TP	↓ cell viability [33]
	SW620	TP	↓ cell viability [24]
		C-PAC	↓ cell proliferation [37]
Esophagus	CP-C	C-PAC	↓ total reactive oxygen species [38]
	JHEsoAD1	C-PAC	↑ autophagy in acid-sensitive cells, pro-death [39,40]
			↑ necrosis in acid-resistant cells [39]
			↑ G_2-M cell cycle arrest [39]
			↑ total reactive oxygen species [38]
			↑ hydrogen peroxide levels [38]
			↓ cell viability [40,41]
			↓ PI3K/AKT/mTOR signaling [39]
	OE33	C-PAC	↑ autophagy in acid-sensitive cells [39]
			↑ low levels of apoptosis [39] ↑ G_2-M cell cycle arrest [39]
			↓ cell proliferation [39]
			↑ total reactive oxygen species [38]
			↓ PI3K/AKT/mTOR signaling [39]
	OE19	C-PAC	↑ necrosis in acid-resistant cells [39]
			↑ G_2-M cell cycle arrest with significant S-phase delay [39]
			↑ total reactive oxygen species [38]
			↑ hydrogen peroxide levels [38]
			↓ PI3K/AKT/mTOR signaling [39]
			↓ cell viability [40,41]
Glioblastoma	SF295	UA	↓ cell density [29]
	U87	C-PAC	↑ apoptosis [36]; ↑ G_1 cell cycle arrest [36]
			↓ cell viability [36]
		Fr6	↑ apoptosis [36]; ↑ G_1 cell cycle arrest [36]
			↓ cell viability [28]
Leukemia	K562	C-PAC	↓ cell density [26]
	RPMI8226	UA	↓ cell density [29]
Liver	HepG2	CE	↑ reduced glutathione levels [22]
			↓ glutathione peroxidase activity [22]
			↓ lipid peroxidation [22]
			↓ reactive oxygen species [22]
		CJE	↑ reduced glutathione levels [22]
			↓ glutathione peroxidase activity [22]
			↓ lipid peroxidation [22]
			↓ reactive oxygen species [22]
		FG	↓ cell viability [27]
		Q	↓ cell viability [27]
		UA	↓ cell viability [27]

Table 1. *Cont.*

Target	Cell Line(s)	Cranberry Constituent	In Vitro Results [Reference(s)]
Lung	DMS114	Fr6	↓ cell viability [28]
	NCI-H322M	UA	↓ cell density [29]
	NCI-H460	C-PAC	↑ apoptosis [37,42]; ↑ G$_1$ cell cycle arrest [37] ↓ cell density [26]; ↓ cell viability [37] ↓ cell proliferation [37]
		UA	↓ cell density [29,30]
Lymphoma	Rev-2-T-6	NDM	↓ cell viability [43] ↓ extracellular matrix invasion [43]
Melanoma	M14	C-PAC UA	↓ cell density [26] ↓ cell density [30]
	SK-MEL5	Fr6	↓ cell viability [28]
Neuroblastoma	IMR-32	C-PAC	↓ cell viability [44]
	SH-Sy5Y	C-PAC	↓ cell viability [44]
	SK-N-SH	C-PAC	↓ cell viability [44]
	SMS-KCNR	C-PAC	↑ apoptosis [44]; ↑ G$_2$-M cell cycle arrest [44] ↑ reactive oxygen species [44] ↓ PI3K/AKT/mTOR signaling [44] ↓ cell viability [44,45]
Oral Cavity	CAL27	CE	↑ apoptosis [46]; ↓ cell adhesion [46] ↓ cell density [46]; ↓ cell viability [24]
		TP	↓ cell viability [33]
	HSC2	CJE	↑ reduced glutathione levels [47]
	KB	CE TP	↓ cell viability [47] ↓ cell viability [24] ↓ cell viability [33]
	SCC25	CE	↑ apoptosis [46]; ↓ cell adhesion [46] ↓ cell density [46]
Ovary	OVCAR-8	C-PAC	↑ G$_2$-M cell cycle arrest [48]; ↓ cell viability [48]
	SKOV-3	C-PAC	↑ apoptosis [48,49]; ↑ G$_2$-M cell cycle arrest [48,49] ↑ reactive oxygen species [49] ↓ AKT signaling [49] ↓ cell proliferation [45,49] ↓ cell viability [45,48,49]
Prostate	22Rv1	CE TP	↓ cell viability [33] ↓ cell viability [33]
	DU-145	CE	↑ G$_1$ cell cycle arrest [50] ↓ cell viability [50,51]
		C-PAC	↑ apoptosis [51] ↑ MAPK signaling [52] ↓ cell viability [26,36,51,52] ↓ matrix metalloprotease activity [52] ↓ PI3K/AKT signaling [52]
		Fr6	↓ cell viability [28,36]
	LNCaP	CE	↓ cell viability [24]
	PC3	CJE	↑ G$_1$ cell cycle arrest [25] ↓ cell viability [25]
		C-PAC UA	↓ cell density [26] ↓ cell density [30]
	RWPE-1	CE	↓ cell viability [33]
		C-PAC TP	↓ cell viability [33] ↓ cell viability [33]
	RWPE-2	CE	↓ cell viability [33]
		C-PAC TP	↓ cell viability [33] ↓ cell viability [33]

Table 1. *Cont.*

Target	Cell Line(s)	Cranberry Constituent	In Vitro Results [Reference(s)]
Renal	RXF393	UA	↓ cell density [29]
	SN12C	UA	↓ cell density [29]
	TK-10	UA	↓ cell density [29]
Stomach	AGS	CJE	↓ cell viability [25]
	SGC-7901	CE	↑ apoptosis [53] ↓ cell proliferation [53] ↓ cell viability [53]

Cranberry derived constituents are abbreviated as follows: anthocyanins (ANTHO), organic-soluble cranberry extract (CE), cranberry juice extract (CJE), cranberry proanthocyanidin-rich fraction (C-PAC), flavonoid-rich fraction 6 (Fr6), flavonoid glycosides (FG), non-dialyzable material from cranberry juice concentrate (NDM), total polyphenolic fraction (TP), quercetin (Q) or ursolic acid fraction (UA). Additional abbreviations: Phosphoinositide 3-kinase (PI3K), Protein Kinase B (AKT), mechanistic Target of Rapamycin (mTOR), mitogen-activated protein kinase (MAPK). Note: MDA-MB-435* was misidentified as a breast cancer cell line, but is now confirmed to be of melanoma origin.

Three cranberry derivatives inhibit the growth of 43 human cancer cell lines. For the purpose of this review, growth inhibitory (GI_{50}) concentrations are presented for cancer cell lines where cranberry treatment results in 50% growth inhibition. Specifically, cranberry derived ursolic acid, proanthocyanidins and an organic-soluble cranberry extract inhibit the growth of breast, colon, cervical, glioblastoma, leukemia, lung, melanoma, oral cavity, prostate and renal cancer cell lines [26,29,30,46]. The GI_{50} concentration for ursolic acid is 1.2–11.2 µM in renal cancer cell lines RXF393, SN12C and TK-10, with similar results observed in breast (MCF-7; 1.4–1.9 µM), colon (HCT116; 1.5–3.5 µM) and lung (NCI-H322M; 1.2–9.8 µM) cancer cell lines when cell density is measured following a 48 h treatment [29,30]. The cell density of lung (NCI-H460; 20.0 µg/mL) and cervical (ME180; 30.0 µg/mL) cancer cell lines are similarly susceptible to cranberry proanthocyanidin treatment, while breast (MCF-7), melanoma (M-14), leukemia (K562) and prostate (PC3) cancer cell lines have GI_{50} values greater than 70.0 µg/mL [26]. An organic soluble cranberry extract is effective at reducing cell density of two oral cancer cells lines, CAL27 and SCC25, at fairly similar GI_{50} concentrations, 40.0 µg/mL and 70.0 µg/mL, respectively [46]. While cell density experiments support a reduction in confluency following treatment with cranberry derivatives, these assays do not take metabolic activity, functional enzyme processes or mechanisms associated with cell death into consideration.

Extending beyond simple density studies, multiple cranberry derived extracts and constituents significantly inhibit the viability of cancer cell lines. In breast cancer cell lines, an organic-soluble cranberry extract, cranberry juice extract, a flavonoid rich fraction (with and without glycosides), quercetin and ursolic acid are all effective at decreasing the viability of MCF-7 and MDA-MB-435 cells [23–25,27,28]. The GI_{50} concentration of a combination flavonoid and glycoside fraction (FG; 23.9 µM) is the most efficacious against MCF-7 cells [27]. Several cranberry extracts inhibit the viability of colon cancer cell lines including an organic-soluble cranberry extract, cranberry juice extract, cranberry proanthocyanidins, a flavonoid rich fraction, a total polyphenolic fraction and ursolic acid [24,25,32,33,35,36]. Cranberry proanthocyanidins and ursolic acid are most effective at inhibiting the viability of HCT116 colon cancer cells, with GI_{50} concentrations of 25 µg/mL for both constituents [35]. LS-513 colon cancer cells are particularly susceptible to viability reductions following exposure to a total polyphenolic fraction (3.8–92.9 µg/mL) and anthocyanins (4.3–75.5 µg/mL) when compared to a cranberry juice extract (38.11–113.0 µg/mL) [32]. Cranberry proanthocyanidins also decrease the viability of neuroblastoma, esophageal adenocarcinoma and ovarian cancer cells [40,41,44,45,48,49]. All four neuroblastoma cancer cell lines (IMR-32, SH-Sy5Y, SK-N-SH and SMS-KCNR) show significant reductions in viability when treated with 12.5–25.0 µg/mL of cranberry proanthocyanidins [44,45]. In comparison, a significant reduction in viability of esophageal adenocarcinoma and ovarian cancer cells is observed with 25.0–50.0 µg/mL and 50.0–200.0 µg/mL cranberry proanthocyanidins, respectively [40,41,45,48,49]. Reductions in viability of glioblastoma and melanoma cancer cell lines occur following treatment with a flavonoid rich extract, with the GI_{50} for U87 glioblastoma cells about half of the GI_{50} for SK-MEL5 melanoma cells after a 96h treatment [28]. A significant decrease in viability is observed for HepG2 liver cancer cells when treated for 4 days with a cranberry flavonoid glycoside extract (GI_{50} = 49.2 µM), quercetin (GI_{50} = 40.9 µM) or ursolic acid (GI_{50} = 87.4 µM); where quercetin was the most effective of the three constituents [27]. Lastly, NCI-H460 cells are more susceptible to a lower dose of cranberry proanthocyanidins (50.0–100.0 µg/mL) than DMS114 cells treated with a flavonoid rich extract; however, both extracts are capable of significantly reducing cell viability of lung cancer cell lines [28,37].

A single study utilized non-dialyzable material isolated from cranberry juice reporting significant reductions in the viability of Rev-T-2-6 lymphoma cells following a 48h treatment [43]. The non-dialyzable material contains high molecular weight polyphenolic compounds, likely containing both proanthocyanidins and smaller quantities of anthocyanins [54]; however, structural characterization was not conducted due to the inability to hydrolyze the high molecular weight components into smaller oligomeric components for MALDI analysis [43]. Cancer cell lines originating from the oral cavity are susceptible to reduced viability following treatment with cranberry extract (50.0–180.6 µg/mL), cranberry juice extract (150.0 µg/mL) and a total polyphenolic fraction (200.0 µg/mL) with cranberry extract (50.0 µg/mL) treatment of CAL27 cells having the lowest GI_{50} [24,33,47]. Ten studies have documented the effectiveness of a cranberry extract, cranberry juice extract, a flavonoid rich extract and cranberry proanthocyanidins at significantly reducing the viability of prostate cancer cells [24–26,28,33,36,45,50–52]. The most efficacious constituent appears to be a cranberry proanthocyanidin extract, which inhibits the viability of RWPE-1 and RWPE-2 prostate cancer cell lines with a GI_{50} of approximately 6.5 µg/mL [33]. In contrast, prostate cancer cells are not particularly sensitive to inhibition by a flavonoid rich extract as the GI_{50} concentration for DU-145 (234.0 µg/mL) cancer cells is comparatively 1.6–3.0 fold higher compared to the GI_{50} for glioblastoma (77.0 µg/mL), melanoma (147.0 µg/mL) and breast (147.0–212.0 µg/mL) cancer cell lines [28]. Finally, two studies report cranberry and cranberry juice extract decrease the viability of stomach cancer cell lines AGS and SGC-7901 [25,53].

The majority of in vitro studies have characterized the ability of cranberry derived extracts and constituents to inhibit cancer cell line density and viability. Cranberry derived proanthocyanidins and ursolic acid tend to have lower GI_{50} values compared to the organic-soluble cranberry extracts and total phenolic fractions. It is difficult to make comparisons to the results observed for the cranberry juice extracts because concentrations are reported as µL/mL and may vary widely in concentrations of inhibitory constituents based on the starting product [25,32,47]. Furthermore, direct comparisons for specific constituents is challenging due to a lack of protocol standardization for the generation and isolation of any cranberry derived extracts or constituents. Taken together, these data show that cranberry derivatives are capable of inhibiting the viability of 31 cancer cell lines.

3.2. Modulation of Cell Proliferation and Cell Cycle Processes by Cranberry Constituents

The ability of cranberry derived extracts and constituents to modulate cell proliferation and cell cycle kinetics is highlighted by studies conducted in nine different target organs. Cell proliferation, as measured by BrdU incorporation, is decreased in colon, lung, ovarian and OE33 esophageal cancer cell lines following treatment with 25.0–125.0 µg/mL cranberry proanthocyanidins [37,39,45,49]. Additional esophageal adenocarcinoma cells treated with cranberry proanthocyanidins responded with an S-phase delay [39], potentially linked to induction of cell death via necrosis. A decrease in cell proliferation of stomach cancer cells is also noted following treatment with a cranberry extract but at a much higher concentration of 10.0 mg/mL [53].

Ten studies present data for modulation of cell cycle progression by cranberry derivatives. In breast cancer cells, an organic-soluble cranberry extract and a flavonoid rich extract induce G_1 and G_2-M cell cycle arrest, respectively [23,28]. In esophageal adenocarcinoma cells, cranberry proanthocyanidins (50.0–100.0 µg/mL) induce G_2-M cell cycle arrest in acid-sensitive JHEsoAd1 and OE33 cells as well as in acid-resistant OE19 cells [39]. However, cranberry proanthocyanidins induce a significant S-phase delay in acid-resistant OE19 cells. In lung cancer cells, cranberry proanthocyanidins (50.0 µg/mL) induce a G_1 cell cycle arrest [37]. Cranberry proanthocyanidins induce a G_2-M cell cycle arrest in neuroblastoma (20.0 µg/mL) and ovarian (75.0–100.0 µg/mL) cancer cells [44,48,49]. Organic soluble cranberry extracts (25.0 µg/mL) and cranberry juice extracts (25.0 µL/mL) are both effective at inducing G_1 cell cycle arrest in prostate cancer cells [25,50]. Finally, cranberry proanthocyanidins are more effective than a flavonoid rich extract at inducing G_1 cell cycle arrest in glioblastoma cells in a time and dose-dependent manner [36].

3.3. Cranberry Derived Extracts and Constituents Induce Cell Death Pathways

Induction of cell death pathways by cranberry constituents has been investigated in nine target organs. An organic soluble cranberry extract induces apoptosis in stomach (5.0 mg/mL) and breast (15.0 mg/mL) cancer cell lines at high concentrations, whereas cancer cell death was induced at much lower concentrations (range of 50.0–70.0 µg/mL) in SCC25 and CAL27 oral cancer cell lines [23,46,53]. Treatment of esophageal adenocarcinoma, lung, colon, glioblastoma, neuroblastoma and ovarian cancer cells with cranberry proanthocyanidins results in apoptosis induction [35–37,39,42,44,48,49]. Ursolic acid (25.0 µg/mL) induces apoptosis in colon cancer cells and a flavonoid rich extract (100.0–400.0 µg/mL) induces late apoptosis in glioblastoma cells [28,35]. Finally, a flavonoid rich extract (200.0–400.0 µg/mL) from the cranberry is also responsible for induction of apoptosis in MDA-MB-435 cancer cells [28].

In esophageal adenocarcinoma cells, the primary form of cell death appears linked to an acid sensitive or acid resistant phenotype. Specifically, treatment with an equimolar concentration of five bile salts (taurocholic, glycocholic, glycodeoxycholic, glycochenodeoxycholic and deoxycholic acids) in acidified medium (pH 4.0) induces rapid apoptosis of esophageal adenocarcinoma cells except in the case of constitutively resistant OE19 cells were exposure to an acidified bile salt mixture has little impact on cell viability. In acid sensitive JHEsoAD1 and OE33 cell lines, cranberry proanthocyanidins induce autophagy through inactivation of Phosphoinositide 3-kinase /Protein Kinase B/mechanistic Target of Rapamycin (PI3K/AKT/mTOR) signaling pathways, but with low levels of apoptosis [39,40]. Conversely, intrinsically acid resistant OE19 cells die mainly through necrosis following treatment with cranberry proanthocyanidins. Acid sensitive JHEsoAd1 cells can be pushed to acid resistance following repeated exposure to an acidified bile cocktail resulting in a switch from autophagic to necrotic cell death [39]. The concentration of cranberry proanthocyanidins necessary to inhibit the viability of acid-resistant cells through necrosis is similar to the concentration inducing autophagy in acid-sensitive cells [39]. It is important to note that autophagy induction in JHEsoAD1 cells is a pro-death mechanism and not a pro-survival response [40,41].

In contrast to esophageal adenocarcinoma cells, the primary form of cell death induced by cranberry proanthocyanidins in neuroblastoma cells is apoptosis [44]. Interestingly, treatment of SMS-KCNR neuroblastoma (25.0 µg/mL) cancer cells for 18 h with cranberry proanthocyanidins also resulted in inactivation of PI3K/AKT/mTOR signaling pathways and at concentrations lower than those necessary for similar effects in esophageal adenocarcinoma (50.0 µg/mL) cancer cells [39,44]. The caspase apoptotic machinery in esophageal adenocarcinoma cells is expressed at low basal levels, which may explain why acid sensitive cells die primarily through autophagy [39]. Furthermore, treatment of acid sensitive esophageal adenocarcinoma cell lines with cranberry proanthocyanidins results in cell death through Beclin-1 independent autophagy induction and is largely caspase-independent despite low levels of apoptosis [39,40]. Conversely, western blot data from untreated neuroblastoma cells show that the apoptotic machinery is present [45]. Thus, cranberry proanthocyanidins modulate cell death via inactivation of the PI3K/AKT/mTOR signaling axis, but may be dependent on available cell death machinery. A decrease in signaling through the AKT pathway is also observed in SKOV-3 ovarian cancer cells treated with cranberry proanthocyanidins [49].

Furthermore, cranberry proanthocyanidins (25.0 µg/mL) increase mitogen-activated protein kinase (MAPK) signaling and decrease PI3K/AKT signaling in prostate cancer cells, suggesting a common cell death mechanism with esophageal adenocarcinoma, ovarian and neuroblastoma cancer cells [52].

3.4. Modulation of Oxidative Status by Cranberries

Consistent with the antioxidant properties of cranberry and cranberry derived extracts in human trials for cardiovascular disease, markers of reactive oxygen species (ROS) decrease in a number of cranberry treated cancer cell lines [15,18,21]. Specifically, malondialdehyde lipid peroxidation levels decrease in colon cancer lines following treatment with a total polyphenolic fraction (250.0 µg/mL) for 24 h [31]. An organic soluble cranberry extract and cranberry juice extract decrease malondialdehyde lipid peroxidation levels in HepG2 liver cancer cells [22]. Furthermore, an organic soluble cranberry extract and cranberry juice extract increase reduced glutathione levels and decrease glutathione peroxidase activity in HepG2 cells, respectively [22]. Total ROS levels also decrease in HepG2 liver cancer cells following treatment with an organic soluble cranberry extract (0.5 µg/mL) and cranberry juice extract (25.0 µg/mL) for 20 h [22].

Conversely, increases in ROS are reported in select cancer cell lines treated with cranberry derivatives. Specifically, cranberry proanthocyanidins increase ROS levels in esophageal adenocarcinoma (100.0 µg/mL), neuroblastoma (20.0 µg/mL) and ovarian (75.0 µg/mL) cancer cell lines [38,44,49]. The increase in JHEsoAD1 esophageal adenocarcinoma cells was partially due to increases in hydrogen peroxide levels [38]. While these data may seem counterintuitive, several drugs including cisplatin, cyclophosphamide and fenretinide used to treat cancer rely on the production of ROS as a mechanism to inhibit cancer cell viability [55,56]. Interestingly, ROS levels decrease in premalignant Barrett's esophageal CP-C cells following treatment with cranberry proanthocyanidins, also resulting in cell death induction [38,57,58]. Given the differences observed between ROS production in premalignant Barrett's and esophageal adenocarcinoma cells following treatment with cranberry proanthocyanidins, it is possible that basal oxidative stress levels or stage of carcinogenesis may inform functional cell death machinery and cell fate. The mechanism of cell death for premalignant esophageal cells treated with cranberry derived constituents remains to be investigated. However, immortalized "normal" Het1A esophageal cells are less susceptible to C-PAC induced cell death compared to esophageal adenocarcinoma and premalignant Barrett's esophageal cell lines when treated with the same cranberry proanthocyanidin fraction at equivalent concentrations [40].

3.5. Additional Biological Processes Modulated by Cranberry Derived Extracts and Constituents

Modulation of pro-inflammatory markers, cellular adhesion, matrix metalloprotease activity and invasion of the extracellular matrix are the remaining biological processes that have been examined in cancer cell lines treated with cranberry based derivatives. Specifically, TNFα and IL-6 levels decrease following treatment with a total cranberry polyphenolic fraction (250.0 µg/mL) in colon cancer cells [31,34]. A soluble cranberry extract (50.0 µg/mL) reduces the expression of cyclooxygenase-2 (COX-2), an enzyme responsible for conversion of arachidonic acid into prostaglandins thus mitigating downstream inflammatory responses [34]. The ability of oral cancer cells to settle, adhere and establish colonies is inhibited by an organic soluble cranberry extract in a dose dependent manner [46]. In prostate cancer cells, cranberry proanthocyanidins (250.0 µg/mL) decrease matrix metalloprotease activity preventing cleavage of extracellular matrix proteins and migration of cancer cells [52]. Finally, a non-dialyzable material extract isolated from cranberry juice (50.0 µg/mL) is capable of decreasing invasion of Rev-2-T-6 lymphoma cells [43]. While these final studies describe distinctly different processes, they provide additional mechanistic information for how cranberry derived extracts and constituents modulate cancer cells in vitro and provide a foundation to support additional in vivo investigations.

3.6. In Vitro Summary

Collectively, data from the in vitro studies support that cranberries successfully inhibit three hallmarks of cancer: resisting cell death, sustaining proliferative signaling and activating invasion and metastasis [59]. These studies provide the preclinical basis for cancer inhibitory investigations of cranberry derived constituents utilizing in vivo models and may inform future clinical trials in high risk human cohorts. Protocol standardization for extraction and isolation of constituents from cranberries will be vital for reproducibility of data and for development of standardized products for use in human clinical trials. Cranberry proanthocyanidins appear to be among the most potent cranberry derived constituents based upon the completed in vitro research evaluating a large panel of cancer cell lines and multiple inhibitory mechanisms. Low concentrations of cranberry derived ursolic acid also favorably impacts cancer cell viability and induces apoptosis, but additional signaling networks have yet to be assessed. However, it is important to recognize that both the starting cranberry product, processing and extraction methods differed greatly across the summarized studies making direct comparisons difficult.

It is interesting to note that included in the in vitro studies is the evaluation of cranberry derived constituents in MDA-MB-435 (misidentified as breast) and M14 melanoma cancer cell lines. Controversy regarding the origin of these two cancer cell lines has been debated since 2003 when microarray studies suggested that the MDA-MB-435 cell line was not of breast origin [60]. Through extensive genetic and expression analyses, it has been established that the MDA-MB-435 cell line is of melanoma origin and is very likely a clone of the M14 melanoma cancer cell line [61]. This finding has broader implications for the scientific community, but herein expands the inhibitory effects of cranberries to include inhibition of melanoma cancer cell viability, modulation of cell cycle kinetics and provides mechanistic insight into mode of cell death induction [25,28,29].

Additional research is necessary for improved understanding of the molecular mechanisms by which cranberry derivatives inhibit cancer cells in vitro including to what extent reactive oxygen species play a role in cell death induction and whether downregulation and inactivation of the PI3K/AKT/mTOR pathway is a common mechanism. The latter is of particular importance for the potential development of neoadjuvant applications combining therapeutic drugs with natural inhibitors [62]. Natural products including quercetin, curcumin, resveratrol and lycopene reportedly impact multiple cancer-associated processes, with clinical trial results supporting the efficacy of lycopene at reducing colon cancer tumor size [63]. In addition, Singh et al. found that cranberry proanthocyanidins acted synergistically with paraplatin to inhibit SKOV-3 ovarian cancer cell viability and proliferation [45]. Importantly, pretreatment of SKOV-3 ovarian cancer cells with cranberry proanthocyanidins significantly reduced the IC$_{50}$ following paraplatin treatment. Utilization of natural products in conjunction with chemotherapy drugs may also be useful in treating patients who have developed resistance to chemotherapeutic drugs [64]. Finally, in esophageal adenocarcinoma JHEsoAD1 cells, pharmacologic inhibition of mTOR using cranberry proanthocyanidins or rapamycin induces autophagy resulting in cell death and survival, respectively [40]. Therefore, understanding the mechanisms of cancer inhibition by cranberry constituents is critical for evaluation of cranberries in human clinical trials with a cancer prevention focus or possibility for use in combination with chemotherapeutic agents.

4. In Vivo Inhibition of Cancer Using Cranberry Products

There are only nine published studies assessing the preclinical efficacy of cranberry products in animal models. These studies are summarized in Table 2 and will be reviewed by target organ in the following section.

Table 2. Summary of preclinical in vivo evaluations of cranberry products as cancer inhibitors.

Target Organ	In Vivo Models/Cranberry Product and Mode of Delivery/Results
[Reference]	
Bladder	
[65]	Nitrosamine-induced tumors in female F344 rats for eight weeks; following a one-week break, treatment with 0.5 mL/rat or 1.0 mL/rat with cranberry juice concentrate by gavage daily for six months; 31% reduction in bladder tumor weight and 38% reduction in cancerous lesion formation.
Colon	
[66]	AOM-induced ACF in male F344 rats three weeks after initiation of cranberry juice treament; ad libitum access to 20% cranberry juice in water for 15 weeks; 77% reduction in AOM-induced ACF with reductions in the proximal and distal colon versus untreated controls; significantly increased levels of liver glutathione-*S*-transferase versus controls.
[36]	HT29 (5.0×10^6 cells) xenografts in female NCR NU/NU mice; treatment with cranberry proanthocyanidins (100.0 mg/kg body weight) intraperitoneally three times weekly for 24 days; significant inhibition of explant growth versus controls.
[67]	DSS induced experimental colitis in male Balb/c mice at weeks three and six; Treatment with cranberry extract powder (0.1% or 1.0%) or 1.5% freeze dried whole cranberry powder in diet ad libitum from start until $\geq$ six weeks; cranberry extract powder (1.0%) and 1.5% dried whole cranberry powder treatment normalized stool consistency, decreased blood in fecal samples versus controls and reduced late onset colitis; all treatments decreased serum TNFα levels.
Esophagus	
[39]	OE19 (1.25×10^6 cells) xenografts in male athymic NU/NU mice; treatment with cranberry proanthocyanidins (250.0 µg/mouse) by oral gavage six days/week for 19 days; 67% decrease in mean tumor volume versus controls and treatment modulated multiple cancer signaling pathways including inactivation of the PI3K/AKT/mTOR pathway.
Glioblastoma	
[36]	U87 (1.0×10^6 cells) xenografts in female NCR NU/NU mice; treatment with cranberry proanthocyanidins (100.0 mg/kg body weight) or a flavonoid rich cranberry fraction (250.0 mg/kg body weight) intraperitoneally three times a week; significant inhibition of explant growth by both fractions versus controls
Lymphoma	
[43]	Rev-2-T-6 (5.0×10^6 cells) xenografts in female Balb/C mice; treatment with non-dialyzable material from cranberry juice concentrate (160.0 mg/kg body weight) intraperitoneally three times a week; significant inhibition of explant growth.
Prostate	
[36]	DU-145 (4.0×10^6 cells) xenografts in female NCR NU/NU mice; treatment with cranberry proanthocyanidins (100.0 mg/kg body weight) intraperitoneally three times a week; significant inhibition of explant growth by cranberry proanthocyanidin fraction.
Stomach	
[53]	SGC-7901 (5.0×10^6 cells) xenografts in Balb/c NU/NU mice; SGC-7901 cells were pre-treated with cranberry extract prior to xenograft implantation; increased tumor latency and reduced tumor size in a dose-dependent manner.

Abbreviations: azoxymethane (AOM), athymic nude mice (NCR NU/NU), dextran sodium sulfate (DSS).

4.1. Cranberry Juice Concentrate and Bladder Cancer

Despite extensive research focused on cranberries and improved urinary tract health, including prevention of urinary tract infections, there is little research on cancer inhibition in vivo. Prasain et al. performed the only in vivo investigation to assess the efficacy of cranberries as inhibitors of bladder cancer [65]. Bladder cancer was induced in female Fischer-F344 rats following administration of *N*-butyl-*N*-(4-hydroxybutyl)-nitrosamine twice a week for eight weeks. Starting one week after the final carcinogen treatment, rats were gavaged daily for six months with cranberry juice concentrate (0.5 mL/rat or 1.0 mL/rat) or water. The cranberry juice concentrate utilized in this study was made available from Ocean Spray Cranberries, Inc. (City, US State, USA) and described to contain 9.57 ± 0.5 mg vanillic acid equivalent/mL of total phenolics. At the end of the six month study, bladder lesions were weighed showing that administration of high dose cranberry juice concentrate (1.0 mL/rat/day) reduced bladder tumor weight by 31%. Both doses of cranberry juice concentrate resulted in a decrease in urinary bladder tumor weight, but only the higher volume of cranberry juice concentrate (1.0 mL/rat/day) resulted in a 38% reduction in cancerous lesion formation in the bladder [65]. These data are promising supporting that an orally administered cranberry juice concentrate is non-toxic when delivered over a period of six months and that a behaviorally achievable concentration significantly inhibits progression events resulting in reduced bladder cancer [15,68].

4.2. Colon Cancer and Cranberries

Two studies have investigated the ability of cranberry products to reduce or inhibit colon carcinogenesis. In the first study, male F344 rats were administered either 20% cranberry juice or water ad libitum for 15 weeks with two weeks of azoxymethane (AOM) administration in weeks 4 and 5 to induce aberrant crypt foci (ACF) and colon cancer [66]. There was a 77% reduction in the number of AOM-induced ACF in rats administered cranberry juice, with reductions in both the proximal and distal colon. Finally, animals had significantly higher levels of liver glutathione-*S*-transferase activity compared to untreated controls supporting that cranberry juice may activate cell protection mechanisms against oxidative stress in the context of colon cancer.

In a study performed by Ferguson et al., the colon carcinoma cell line HT-29 was utilized to establish xenografts in female NCR NU/NU mice [36]. Specifically, HT-29 (5.0×10^6) cells were subcutaneously injected into the right flank of mice and tumors were monitored every other day. Cranberry proanthocyanidins (100.0 mg/kg) were administered intraperitoneally 2–3 times per week for a total of ten injections over 24 days. Mice treated with cranberry proanthocyanidins had significantly lower body weights early in the study but weights returned to normal in the last five days of the study raising the question of initial toxicity at the administered concentration. Importantly, cranberry proanthocyanidins significantly inhibited the growth of HT-29 tumor xenografts in mice.

The final colon related study utilized a dextran sodium sulfated (DSS)-induced mouse model of colitis [67]. DSS-induced colitis is widely accepted as an animal model of irritable bowel syndrome, a condition linked to elevated risk for colorectal cancer development [69]. In this study, male Balb/C mice were fed a diet containing either cranberry extract (0.1% or 1%) or 1.5% dried whole cranberry powder in the food *ad libitum*. On the third and sixth weeks of the study, the water was replaced with 1% w/v of DSS solution for one week. Animals fed either the cranberry extract or the dried whole cranberry powder had a significantly lower disease activity index (more normal stool consistency and decreased blood in fecal samples) compared to controls for both DSS treatment periods. Furthermore, 1% cranberry extract and the 1.5% dried whole cranberry powder diets delayed the onset of colitis. In addition, inflammatory markers including TNFα (significant in all cranberry treatments) and IL-1β (1.5% dried whole cranberry powder only) were decreased in colonic tissues, suggesting that cranberry derived extracts and powders exert an anti-inflammatory role in vivo.

These data parallel in vitro data for cell death induction, antioxidant effects and anti-inflammatory properties of cranberries and cranberry derived constituents. In the studies by Boateng et al. and Xiao et al. [66,67], the cranberry products were administered throughout the entire bioassay and although encouraging consideration in future studies should be given to discerning whether cranberry products possess both anti-initiation as well as anti-promotion/progression properties in vivo. As colon cancer and a western diet are reportedly linked, regular inclusion of cranberries may impart cancer preventative advantages [70].

4.3. Cranberry Proanthocyanidins and Esophageal Adenocarcinoma

The cancer inhibitory potential of cranberry proanthocyanidins against esophageal adenocarcinoma was investigated by Kresty et al. using murine xenografts [39]. Acid-resistant OE19 (1.25×10^6) cells were subcutaneously implanted in each flank of twelve male NU/NU athymic mice and tumors were established (150 mm^3) prior to initiation of cranberry proanthocyanidin treatment. Five days after cell injection, mice were randomized for treatment with cranberry proanthocyanidins (250.0 µg/mouse) or vehicle by oral gavage six days a week. There was a significant difference in tumor volume between vehicle and cranberry proanthocyanidin treated mice by day 12, with the study ending on day 19 due to large tumor size among vehicle treated controls. Mean tumor volume in cranberry proanthocyanidin treated mice was reduced by 67.6% and showed reduced inflammation compared to tumors isolated from control mice. Importantly, tumor lysates from cranberry proanthocyanidin treated mice showed inactivation of PI3K/AKT/mTOR signaling as was observed in vitro utilizing a panel of esophageal adenocarcinoma cell lines [39]. Interestingly, acid-sensitive JHAD1 and OE33 cells were not able to form tumors in this xenograft model, suggesting that the OE19 cells are phenotypically more aggressive and that acid-resistance may support tumor development.

4.4. Glioblastoma and Cranberry Derived Constituents

The efficacy of cranberry proanthocyanidins and a flavonoid rich cranberry extract for prevention of glioblastoma tumors was evaluated by Ferguson et al. in a murine xenograft model [36]. U87 (1.0×10^6) glioblastoma multiforme cancer cells were injected subcutaneously into the right flank of female NCR NU/NU mice and tumors were monitored every other day. Mice were administered cranberry proanthocyanidins (100.0 mg/kg body weight) or a flavonoid rich extract (250.0 mg/kg body weight) intraperitoneally 2–3 times per week for a total of ten injections. Both cranberry fractions were able to slow tumor growth by up to 40%. These data suggest that cranberry derived extracts may be effective against glioblastoma muliforme but the mechanisms remain to be characterized.

4.5. Non-Dialyzable Material from Cranberry Juice and Lymphoma

Inhibition of lymphoma xenograft growth by cranberry juice in a murine model was investigated by Hochman et al. [43]. Rev-2-T-6 (5.0×10^6) lymphoma cancer cells were injected into female Balb/C mice and animals were subsequently treated with cranberry juice constituents (130.0 mg/kg body weight) intraperitoneally every other day for two weeks. The treatment consisted of non-dialyzable material from cranberry juice with a molecular weight range of 12–30 kDa. Mice treated with non-dialyzable material did not form tumors with 80% of control mice developing lymphomas. This study is positive for the prevention of lymphoma by cranberry juice but due to study design the results fail to discriminate between failed xenograft implementation and prevention of tumor growth.

4.6. Prostate Cancer and Cranberry Proanthocyanidins

Utilizing a murine xenograft model with prostate cancer cells, Ferguson et al. showed that cranberry proanthocyanidins inhibit tumors in vivo [36]. DU-145 (4.0×10^6) prostate cancer cells were subcutaneously implanted in the right flank of female NCR NU/NU mice. On day 2, mice were administered cranberry proanthocyanidins (100.0 mg/kg body weight) intraperitoneally every 2 or 3 days for a total of 10 injections. Cranberry proanthocyanidins significantly inhibited the growth of tumors compared to control mice and two of the mice had tumor regression by 108 days post implant. These data are encouraging for the inhibition of prostate tumor development by cranberry proanthocyanidins, but additional experiments are necessary for investigating the mechanism of tumor regression.

4.7. Whole Cranberry Extract and Stomach Cancer

Efficacy of a whole cranberry extract inhibiting stomach cancer tumor growth was evaluated in a murine xenograft model developed by Liu et al. [53]. Balb/C NU/NU mice were subcutaneously injected with SGC-7901 (5.0×10^6) gastric adenocarcinoma cells in the right flank region. Prior to implantation, SGC-7901 cells were pretreated with a whole cranberry extract (0–40.0 mg/mL) for 48 h. Tumors in control mice developed by the tenth day of the experiment with a delay of 3–7 days for cells pretreated with 5.0–20.0 mg/mL of cranberry extract. Xenograft tumors did not develop from SGC-7901 cells pretreated with 40.0 mg/mL cranberry extract but in vitro data presented in the same study supported that these cells were likely undergoing high levels of apoptosis with reduced viability. Therefore, additional studies will be necessary to determine how a whole cranberry extract inhibits the growth of stomach cancer in animal models, including those which model gastric carcinogenesis as a result of chronic *Helicobacter pylori* infection [71].

4.8. In Vivo Summary

The in vivo studies described here provide preliminary evidence for the preclinical efficacy of multiple cranberry derived extracts against seven cancer targets. Except for the studies in esophageal adenocarcinoma and colon cancer, the majority of completed in vivo studies reporting inhibition of tumor development or growth, fail to include further mechanistic assessments. Of the nine in vivo studies, three studies used carcinogens or chemicals to induce cancer in animal models. In the bladder, delivery of a cranberry juice concentrate by gavage following carcinogen treatment supports anti-promotion/progression effects of cranberries against chemically-induced bladder cancer. Two studies in the colon assessed the efficacy of cranberry juice, cranberry extract powder and a dried whole cranberry powder in a full carcinogenesis schematic, where dietary administration of cranberries began prior to carcinogen initiation and continued throughout, after carcinogen or chemical treatment was completed. In regard to mode of delivery, four of the in vivo studies delivered the cranberry product by orally, either in water, diet or gavage with efficacy suggesting that the compounds or their metabolites hold promise as orally bioavailable cancer inhibitors. Administration of cranberry products via intraperitoneal injection also showed cancer inhibitory efficacy in four in vivo studies, but as a mode of delivery is less relevant for primary or secondary cancer prevention efforts in human cohorts. Overall, the in vivo results expand upon in vitro observations and importantly support that long-term administration of cranberry products is well tolerated and cancer inhibitory in various animal models. However, additional research focused on bioavailability, metabolic fate and additional cancer inhibitory mechanisms of cranberry products is warranted for informing clinical focused cancer prevention efforts.

To date only a few human studies have characterized cranberry metabolites in plasma or urine and often these studies are limited to quantifying molecules that have previously been identified [15,16,72–75]. A recent study by McKay et al. reported that flavonoids, phenolic acids and proanthocyanidins can be detected in the urine or plasma of individuals who consumed a 54% cranberry juice cocktail [15]. Recent advances in standards used for identification and quantification of cranberry metabolites has resulted in the identification of 60 metabolites from the urine and plasma of healthy men after consumption of a cranberry juice cocktail that contained 787 mg of polyphenols [68]. The ability to detect and quantitate proanthocyanidins in the urine and plasma is not consistent from study to study, but this should improve with the recent development of a new cranberry proanthocyanidin standard that more closely reflects the structural heterogeneity of proanthocyanidins present in fresh cranberries [76]. Cranberry proanthocyanidins are large, complex molecules with recent data supporting that the intestinal microbiome is responsible for the metabolism of cranberry proanthocyanidins into smaller active metabolites [68]. Additional research will be necessary to assess the bioavailability and metabolism of cranberries in humans and recent advances in standards and the radiolabeling of cranberry products will provide new tools to aid investigations.

5. Conclusions

Evaluation of cranberries and cranberry derived constituents in preclinical in vitro and in vivo studies evaluating cancer inhibition is key for the future development of cranberry-based interventions in high-risk human cohorts. The data presented in this review strongly support the anti-proliferative and pro-death capacities of cranberries in a multitude of cancer cell lines and select in vivo models including xenograft and chemically induced cancer models. The precise cancer inhibitory mechanisms associated with cranberries in specific targets are still be elucidated, but preclinical studies utilizing cranberry proanthocyanidins show inactivation of the PI3K/AKT/mTOR pathways and modulation of MAPK signaling in esophageal, neuroblastoma, ovarian and prostate cancer cells and in esophageal xenografts [39,44,49,52]. Moreover, cranberry proanthocyanidins have recently been shown to induce autophagic markers in vitro and in vivo [39], suggesting an alternative mode of cell death induction and tumor inhibition that requires further evaluation in additional cancer targets. A recent study published by The Cancer Genome Atlas Network showed that a large number of genetic alterations were shared across 279 patients with head and neck squamous cell carcinomas including activating mutations in *PIK3CA*, the gene encoding the catalytic subunit of phosphatidylinositol 3-kinase (PI3K) [77]. Cranberry proanthocyanidins are also known to mitigate inflammatory responses of oral epithelial cells and to inhibit oral biofilm formation [12,78]; thus, cranberry derived constituents may be particularly efficacious inhibitors targeting oral premalignancy. Researchers should continue to define the mechanisms of cancer inhibition in vitro and in vivo with the goal of informing mechanistically driven human clinical trials. It should be noted that the efficacy of natural products in head and neck, esophageal and colon cancers has been demonstrated by black raspberries [79]. The use of cranberries and cranberry derived constituents in cancer prevention is at an early stage. Still, results are highly promising considering positive preclinical results following treatment at relatively low, behaviorally achievable, concentrations when administered in a drink formulation, consumed as food or as a supplement. A recent study by Marette *et al.* reported cranberry polyphenols protect from diet-induced obesity, insulin resistance and intestinal inflammation [80]. The latter research findings were associated with a shift in fecal microbiome profiles and although cancer was not an outcome under evaluation, further assessments of cranberries targeting obesity-linked cancers seems logical. In conclusion, additional research focused on issues of metabolism, bioavailability, pharmacokinetics, pharmacodynamics, active fractionation, optimum dose, formulation, routes of delivery and duration are required to inform the cancer preventive utility of cranberries in high risk human cohorts.

Acknowledgments: We thank the National Institutes of Health and National Cancer Institute for funding support via grant (R01-CA158319) and Advancing a Healthier Wisconsin (5220296), both awarded to Laura A. Kresty.

Author Contributions: Katherine M. Weh conducted analysis of data and drafting of the paper; Jennifer Clarke assessed study methodology and statistical approaches; Laura A. Kresty completed analysis of data, drafting of the paper and final approval of the version for publication.

Conflicts of Interest: The authors declare no conflict of interest.

References

1. Blumberg, J.B.; Camesano, T.A.; Cassidy, A.; Kris-Etherton, P.; Howell, A.; Manach, C.; Ostertag, L.M.; Sies, H.; Skulas-Ray, A.; Vita, J.A. Cranberries and their bioactive constituents in human health. *Adv. Nutr.* **2013**, *4*, 618–632. [CrossRef] [PubMed]
2. Neto, C.C. Cranberry and blueberry: Evidence for protective effects against cancer and vascular diseases. *Mol. Nutr. Food Res.* **2007**, *51*, 652–664. [CrossRef] [PubMed]

3. Nantz, M.P.; Rowe, C.A.; Muller, C.; Creasy, R.; Colee, J.; Khoo, C.; Percival, S.S. Consumption of cranberry polyphenols enhances human gamma delta-T cell proliferation and reduces the number of symptoms associated with colds and influenza: A randomized, placebo-controlled intervention study. *Nutr. J.* **2013**, *12*, 161. [CrossRef] [PubMed]

4. Steinberg, D.; Feldman, M.; Ofek, I.; Weiss, E.I. Cranberry high molecular weight constituents promote streptococcus sobrinus desorption from artificial biofilm. *Int. J. Antimicrob. Agents* **2005**, *25*, 247–251. [CrossRef] [PubMed]

5. Howell, A.B.; Foxman, B. Cranberry juice and adhesion of antibiotic-resistant uropathogens. *JAMA* **2002**, *287*, 3082–3083. [CrossRef] [PubMed]

6. Howell, A.B.; Vorsa, N.; Der Marderosian, A.; Foo, L.Y. Inhibition of the adherence of p-fimbriated *Escherichia coli* to uroepithelial-cell surfaces by proanthocyanidin extracts from cranberries. *N. Engl. J. Med.* **1998**, *339*, 1085–1086. [CrossRef] [PubMed]

7. Howell, A.; Souza, D.; Roller, M.; Fromentin, E. Comparison of the anti-adhesion activity of three different cranberry extracts on uropathogenic p-fimbriated *Escherichia coli*: A randomized, double-blind, placebo controlled, ex vivo, acute study. *Nat. Prod. Commun.* **2015**, *10*, 1215–1218. [PubMed]

8. Howell, A.B.; Botto, H.; Combescure, C.; Blanc-Potard, A.B.; Gausa, L.; Matsumoto, T.; Tenke, P.; Sotto, A.; Lavigne, J.P. Dosage effect on uropathogenic *Escherichia coli* anti-adhesion activity in urine following consumption of cranberry powder standardized for proanthocyanidin content: A multicentric randomized double blind study. *BMC Infect. Dis.* **2010**, *10*, 94. [CrossRef] [PubMed]

9. Howell, A.B.; Reed, J.D.; Krueger, C.G.; Winterbottom, R.; Cunningham, D.G.; Leahy, M. A-type cranberry proanthocyanidins and uropathogenic bacterial anti-adhesion activity. *Phytochemistry* **2005**, *66*, 2281–2291. [CrossRef] [PubMed]

10. Kaspar, K.L.; Howell, A.B.; Khoo, C. A randomized, double-blind, placebo-controlled trial to assess the bacterial anti-adhesion effects of cranberry extract beverages. *Food Funct.* **2015**, *6*, 1212–1217. [CrossRef] [PubMed]

11. Shmuely, H.; Yahav, J.; Samra, Z.; Chodick, G.; Koren, R.; Niv, Y.; Ofek, I. Effect of cranberry juice on eradication of *Helicobacter pylori* in patients treated with antibiotics and a proton pump inhibitor. *Mol. Nutr. Food Res.* **2007**, *51*, 746–751. [CrossRef] [PubMed]

12. Girardot, M.; Guerineau, A.; Boudesocque, L.; Costa, D.; Bazinet, L.; Enguehard-Gueiffier, C.; Imbert, C. Promising results of cranberry in the prevention of oral Candida biofilms. *Pathog. Dis.* **2014**, *70*, 432–439. [CrossRef] [PubMed]

13. Lee, Y.L.; Owens, J.; Thrupp, L.; Barron, S.; Shanbrom, E.; Cesario, T.; Najm, W.I. The antifungal activity of urine after ingestion of cranberry products. *J. Altern. Complement. Med.* **2009**, *15*, 957–958. [CrossRef] [PubMed]

14. Pappas, E.; Schaich, K.M. Phytochemicals of cranberries and cranberry products: Characterization, potential health effects, and processing stability. *Crit. Rev. Food Sci. Nutr.* **2009**, *49*, 741–781. [CrossRef] [PubMed]

15. McKay, D.L.; Chen, C.Y.; Zampariello, C.A.; Blumberg, J.B. Flavonoids and phenolic acids from cranberry juice are bioavailable and bioactive in healthy older adults. *Food Chem.* **2015**, *168*, 233–240. [CrossRef] [PubMed]

16. Milbury, P.E.; Vita, J.A.; Blumberg, J.B. Anthocyanins are bioavailable in humans following an acute dose of cranberry juice. *J. Nutr.* **2010**, *140*, 1099–1104. [CrossRef] [PubMed]

17. Walsh, J.M.; Ren, X.; Zampariello, C.; Polasky, D.A.; McKay, D.L.; Blumberg, J.B.; Chen, C.Y. Liquid chromatography with tandem mass spectrometry quantification of urinary proanthocyanin a2 dimer and its potential use as a biomarker of cranberry intake. *J. Sep. Sci.* **2016**, *39*, 342–349. [CrossRef] [PubMed]

18. Mathison, B.D.; Kimble, L.L.; Kaspar, K.L.; Khoo, C.; Chew, B.P. Consumption of cranberry beverage improved endogenous antioxidant status and protected against bacteria adhesion in healthy humans: A randomized controlled trial. *Nutr. Res.* **2014**, *34*, 420–427. [CrossRef] [PubMed]

19. Novotny, J.A.; Baer, D.J.; Khoo, C.; Gebauer, S.K.; Charron, C.S. Cranberry juice consumption lowers markers of cardiometabolic risk, including blood pressure and circulating c-reactive protein, triglyceride, and glucose concentrations in adults. *J. Nutr.* **2015**, *145*, 1185–1193. [CrossRef] [PubMed]

20. Rodriguez-Mateos, A.; Feliciano, R.P.; Boeres, A.; Weber, T.; Dos Santos, C.N.; Ventura, M.R.; Heiss, C. Cranberry (poly)phenol metabolites correlate with improvements in vascular function: A double-blind, randomized, controlled, dose-response, crossover study. *Mol. Nutr. Food Res.* **2016**. in press. [CrossRef] [PubMed]

21. Valentova, K.; Stejskal, D.; Bednar, P.; Vostalova, J.; Cihalik, C.; Vecerova, R.; Koukalova, D.; Kolar, M.; Reichenbach, R.; Sknouril, L.; et al. Biosafety, antioxidant status, and metabolites in urine after consumption of dried cranberry juice in healthy women: A pilot double-blind placebo-controlled trial. *J. Agric. Food Chem.* **2007**, *55*, 3217–3224. [CrossRef] [PubMed]

22. Martin, M.A.; Ramos, S.; Mateos, R.; Marais, J.P.J.; Bravo-Clemente, L.; Khoo, C.; Goya, L. Chemical characterization and chemo-protective activity of cranberry phenolic powders in a model cell culture. Response of the antioxidant defenses and regulation of signaling pathways. *Food Res. Int.* **2015**, *71*, 68–82. [CrossRef]

23. Sun, J.; Liu, R.H. Cranberry phytochemical extracts induce cell cycle arrest and apoptosis in human MCF-7 breast cancer cells. *Cancer Lett.* **2006**, *241*, 124–134. [CrossRef] [PubMed]

24. Seeram, N.P.; Adams, L.S.; Zhang, Y.J.; Lee, R.; Sand, D.; Scheuller, H.S.; Heber, D. Blackberry, black raspberry, blueberry, cranberry, red raspberry, and strawberry extracts inhibit growth and stimulate apoptosis of human cancer cells in vitro. *J. Agric. Food Chem.* **2006**, *54*, 9329–9339. [CrossRef] [PubMed]

25. Boivin, D.; Blanchette, M.; Barrette, S.; Moghrabi, A.; Beliveau, R. Inhibition of cancer cell proliferation and suppression of TNF-induced activation of NFKB by edible berry juice. *Anticancer Res.* **2007**, *27*, 937–948. [PubMed]

26. Neto, C.C.; Krueger, C.G.; Lamoureaux, T.L.; Kondo, M.; Vaisberg, A.J.; Hurta, R.A.R.; Curtis, S.; Matchett, M.D.; Yeung, H.; Sweeney, M.I.; et al. Maldi-TOF MS characterization of proanthocyanidins from cranberry fruit (*Vaccinium macrocarpon*) that inhibit tumor cell growth and matrix metalloproteinase expression in vitro. *J. Sci. Food. Agr.* **2006**, *86*, 18–25. [CrossRef]

27. He, X.J.; Liu, R.H. Cranberry phytochemicals: Isolation, structure elucidation, and their antiproliferative and antioxidant activities. *J. Agric. Food Chem.* **2006**, *54*, 7069–7074. [CrossRef] [PubMed]

28. Ferguson, P.J.; Kurowska, E.; Freeman, D.J.; Chambers, A.F.; Koropatnick, D.J. A flavonoid fraction from cranberry extract inhibits proliferation of human tumor cell lines. *J. Nutr.* **2004**, *134*, 1529–1535. [PubMed]

29. Kondo, M.; MacKinnon, S.L.; Craft, C.C.; Matchett, M.D.; Hurta, R.A.; Neto, C.C. Ursolic acid and its esters: Occurrence in cranberries and other vaccinium fruit and effects on matrix metalloproteinase activity in DU145 prostate tumor cells. *J. Sci. Food Agric.* **2011**, *91*, 789–796. [CrossRef] [PubMed]

30. Murphy, B.T.; MacKinnon, S.L.; Yan, X.J.; Hammond, G.B.; Vaisberg, A.J.; Neto, C.C. Identification of triterpene hydroxycinnamates with in vitro antitumor activity from whole cranberry fruit (*Vaccinium macrocarpon*). *J. Agric. Food Chem.* **2003**, *51*, 3541–3545. [CrossRef] [PubMed]

31. Denis, M.C.; Desjardins, Y.; Furtos, A.; Marcil, V.; Dudonne, S.; Montoudis, A.; Garofalo, C.; Delvin, E.; Marette, A.; Levy, E. Prevention of oxidative stress, inflammation and mitochondrial dysfunction in the intestine by different cranberry phenolic fractions. *Clin. Sci. (Lond.)* **2015**, *128*, 197–212. [CrossRef] [PubMed]

32. Vu, K.D.; Carlettini, H.; Bouvet, J.; Cote, J.; Doyon, G.; Sylvain, J.F.; Lacroix, M. Effect of different cranberry extracts and juices during cranberry juice processing on the antiproliferative activity against two colon cancer cell lines. *Food Chem.* **2012**, *132*, 959–967. [CrossRef]

33. Seeram, N.P.; Adams, L.S.; Hardy, M.L.; Heber, D. Total cranberry extract versus its phytochemical constituents: Antiproliferative and synergistic effects against human tumor cell lines. *J. Agric. Food Chem.* **2004**, *52*, 2512–2517. [CrossRef] [PubMed]

34. Narayansingh, R.; Hurta, R.A.R. Cranberry extract and quercetin modulate the expression of cyclooxygenase-2 (COX-2) and IKBA in human colon cancer cells. *J. Sci. Food Agric.* **2009**, *89*, 542–547. [CrossRef]

35. Liberty, A.M.; Neto, C. Cranberry PACs and triterpenoids: Anti-cancer activities in colon tumor cell lines. *Acta Hortic.* **2009**, 61–66. [CrossRef]

36. Ferguson, P.J.; Kurowska, E.M.; Freeman, D.J.; Chambers, A.F.; Koropatnick, J. In vivo inhibition of growth of human tumor lines by flavonoid fractions from cranberry extract. *Nutr. Cancer* **2006**, *56*, 86–94. [CrossRef] [PubMed]

37. Kresty, L.A.; Howell, A.B.; Baird, M. Cranberry proanthocyanidins induce apoptosis and inhibit acid-induced proliferation of human esophageal adenocarcinoma cells. *J. Agric. Food Chem.* **2008**, *56*, 676–680. [CrossRef] [PubMed]

38. Weh, K.M.; Aiyer, H.S.; Howell, A.B.; Kresty, L.A. Cranberry proanthocyanidins modulate reactive oxygen species in Barrett's and esophageal adenocarcinoma cell lines. *J. Berry Res.* **2016**, *6*, 125–136. [CrossRef]

39. Kresty, L.A.; Weh, K.M.; Zeyzus-Johns, B.; Perez, L.N.; Howell, A.B. Cranberry proanthocyanidins inhibit esophageal adenocarcinoma in vitro and in vivo through pleiotropic cell death induction and PI3K/AKT/mTOR inactivation. *Oncotarget* **2015**, *6*, 33438–33455. [PubMed]

40. Weh, K.M.; Howell, A.; Kresty, L.A. Expression, modulation, and clinical correlates of the autophagy protein Beclin-1 in esophageal adenocarcinoma. *Mol. Carcinog.* **2015**. in press. [CrossRef]

41. Kresty, L.A.; Clarke, J.; Ezell, K.; Exum, A.; Howell, A.B.; Guettouche, T. MicroRNA alterations in Barrett's esophagus, esophageal adenocarcinoma, and esophageal adenocarcinoma cell lines following cranberry extract treatment: Insights for chemoprevention. *J. Carcinog.* **2011**, *10*, 34. [CrossRef] [PubMed]

42. Kresty, L.A.; Howell, A.B.; Baird, M. Cranberry proanthocyanidins mediate growth arrest of lung cancer cells through modulation of gene expression and rapid induction of apoptosis. *Molecules* **2011**, *16*, 2375–2390. [CrossRef] [PubMed]

43. Hochman, N.; Houri-Haddad, Y.; Koblinski, J.; Wahl, L.; Roniger, M.; Bar-Sinai, A.; Weiss, E.I.; Hochman, J. Cranberry juice constituents impair lymphoma growth and augment the generation of antilymphoma antibodies in syngeneic mice. *Nutr. Cancer* **2008**, *60*, 511–517. [CrossRef] [PubMed]

44. Singh, A.P.; Lange, T.S.; Kim, K.K.; Brard, L.; Horan, T.; Moore, R.G.; Vorsa, N.; Singh, R.K. Purified cranberry proanthocyanidines (PAC-1A) cause pro-apoptotic signaling, ROS generation, cyclophosphamide retention and cytotoxicity in high-risk neuroblastoma cells. *Int. J. Oncol.* **2012**, *40*, 99–108. [PubMed]

45. Singh, A.P.; Singh, R.K.; Kim, K.K.; Satyan, K.S.; Nussbaum, R.; Torres, M.; Brard, L.; Vorsa, N. Cranberry proanthocyanidins are cytotoxic to human cancer cells and sensitize platinum-resistant ovarian cancer cells to paraplatin. *Phytother. Res.* **2009**, *23*, 1066–1074. [CrossRef] [PubMed]

46. Chatelain, K.; Phippen, S.; McCabe, J.; Teeters, C.A.; O'Malley, S.; Kingsley, K. Cranberry and grape seed extracts inhibit the proliferative phenotype of oral squamous cell carcinomas. *Evid. Based Complement. Altern. Med.* **2011**, 1–12. [CrossRef] [PubMed]

47. Babich, H.; Ickow, I.M.; Weisburg, J.H.; Zuckerbraun, H.L.; Schuck, A.G. Cranberry juice extract, a mild prooxidant with cytotoxic properties independent of reactive oxygen species. *Phytother. Res.* **2012**, *26*, 1358–1365. [CrossRef] [PubMed]

48. Wang, Y.; Han, A.; Chen, E.; Singh, R.K.; Chichester, C.O.; Moore, R.G.; Singh, A.P.; Vorsa, N. The cranberry flavonoids PAC DP-9 and quercetin aglycone induce cytotoxicity and cell cycle arrest and increase cisplatin sensitivity in ovarian cancer cells. *Int. J. Oncol.* **2015**, *46*, 1924–1934. [PubMed]

49. Kim, K.K.; Singh, A.P.; Singh, R.K.; Demartino, A.; Brard, L.; Vorsa, N.; Lange, T.S.; Moore, R.G. Anti-angiogenic activity of cranberry proanthocyanidins and cytotoxic properties in ovarian cancer cells. *Int. J. Oncol.* **2012**, *40*, 227–235. [CrossRef] [PubMed]

50. Deziel, B.; MacPhee, J.; Patel, K.; Catalli, A.; Kulka, M.; Neto, C.; Gottschall-Pass, K.; Hurta, R. American cranberry (*Vaccinium macrocarpon*) extract affects human prostate cancer cell growth via cell cycle arrest by modulating expression of cell cycle regulators. *Food Funct.* **2012**, *3*, 556–564. [CrossRef] [PubMed]

51. MacLean, M.A.; Scott, B.E.; Deziel, B.A.; Nunnelley, M.C.; Liberty, A.M.; Gottschall-Pass, K.T.; Neto, C.C.; Hurta, R.A. North American cranberry (*Vaccinium macrocarpon*) stimulates apoptotic pathways in DU145 human prostate cancer cells in vitro. *Nutr. Cancer* **2011**, *63*, 109–120. [CrossRef] [PubMed]

52. Deziel, B.A.; Patel, K.; Neto, C.; Gottschall-Pass, K.; Hurta, R.A. Proanthocyanidins from the American cranberry (*Vaccinium macrocarpon*) inhibit matrix metalloproteinase-2 and matrix metalloproteinase-9 activity in human prostate cancer cells via alterations in multiple cellular signalling pathways. *J. Cell Biochem.* **2010**, *111*, 742–754. [CrossRef] [PubMed]

53. Liu, M.; Lin, L.Q.; Song, B.B.; Wang, L.F.; Zhang, C.P.; Zhao, J.L.; Liu, J.R. Cranberry phytochemical extract inhibits SGC-7901 cell growth and human tumor xenografts in Balb/c NU/NU mice. *J. Agric. Food Chem.* **2009**, *57*, 762–768. [CrossRef] [PubMed]

54. Bodet, C.; Chandad, F.; Grenier, D. Anti-inflammatory activity of a high-molecular-weight cranberry fraction on macrophages stimulated by lipopolysaccharides from periodontopathogens. *J. Dent. Res.* **2006**, *85*, 235–239. [CrossRef] [PubMed]

55. O'Donovan, T.R.; O'Sullivan, G.C.; McKenna, S.L. Induction of autophagy by drug-resistant esophageal cancer cells promotes their survival and recovery following treatment with chemotherapeutics. *Autophagy* **2011**, *7*, 509–524. [CrossRef] [PubMed]

56. Osone, S.; Hosoi, H.; Kuwahara, Y.; Matsumoto, Y.; Iehara, T.; Sugimoto, T. Fenretinide induces sustained-activation of JNK/p38 MAPK and apoptosis in a reactive oxygen species-dependent manner in neuroblastoma cells. *Int. J. Cancer* **2004**, *112*, 219–224. [CrossRef] [PubMed]

57. Kavanagh, M.E.; O'Sullivan, K.E.; O'Hanlon, C.; O'Sullivan, J.N.; Lysaght, J.; Reynolds, J.V. The esophagitis to adenocarcinoma sequence; the role of inflammation. *Cancer Lett.* **2014**, *345*, 182–189. [CrossRef] [PubMed]

58. Farhadi, A.; Fields, J.; Banan, A.; Keshavarzian, A. Reactive oxygen species: Are they involved in the pathogenesis of GERD, Barrett's esophagus, and the latter's progression toward esophageal cancer? *Am. J. Gastroenterol.* **2002**, *97*, 22–26. [CrossRef] [PubMed]

59. Hanahan, D.; Weinberg, R.A. Hallmarks of cancer: The next generation. *Cell* **2011**, *144*, 646–674. [CrossRef] [PubMed]

60. Paz, M.F.; Fraga, M.F.; Avila, S.; Guo, M.; Pollan, M.; Herman, J.G.; Esteller, M. A systematic profile of DNA methylation in human cancer cell lines. *Cancer Res.* **2003**, *63*, 1114–1121. [PubMed]

61. Prasad, V.V.T.S.; Gopalan, R.O.G. Continued use of MDA-MB-435, a melanoma cell line, as a model for human breast cancer, even in year, 2014. *NPJ Breast Cancer* **2015**, *1*, 15002. [CrossRef]

62. Agbarya, A.; Ruimi, N.; Epelbaum, R.; Ben-Arye, E.; Mahajna, J. Natural products as potential cancer therapy enhancers: A preclinical update. *SAGE Open Med.* **2014**, *2*. [CrossRef] [PubMed]

63. HemaIswarya, S.; Doble, M. Potential synergism of natural products in the treatment of cancer. *Phytother. Res.* **2006**, *20*, 239–249. [CrossRef] [PubMed]

64. Wu, C.P.; Ohnuma, S.; Ambudkar, S.V. Discovering natural product modulators to overcome multidrug resistance in cancer chemotherapy. *Curr. Pharm. Biotechnol.* **2011**, *12*, 609–620. [CrossRef] [PubMed]

65. Prasain, J.K.; Jones, K.; Moore, R.; Barnes, S.; Leahy, M.; Roderick, R.; Juliana, M.M.; Grubbs, C.J. Effect of cranberry juice concentrate on chemically-induced urinary bladder cancers. *Oncol. Rep.* **2008**, *19*, 1565–1570. [PubMed]

66. Boateng, J.; Verghese, M.; Shackelford, L.; Walker, L.T.; Khatiwada, J.; Ogutu, S.; Williams, D.S.; Jones, J.; Guyton, M.; Asiamah, D.; et al. Selected fruits reduce azoxymethane (AOM)-induced aberrant crypt foci (ACF) in Fisher 344 male rats. *Food Chem. Toxicol.* **2007**, *45*, 725–732. [CrossRef] [PubMed]

67. Xiao, X.; Kim, J.; Sun, Q.; Kim, D.; Park, C.S.; Lu, T.S.; Park, Y. Preventive effects of cranberry products on experimental colitis induced by dextran sulphate sodium in mice. *Food Chem.* **2015**, *167*, 438–446. [CrossRef] [PubMed]

68. Feliciano, R.P.; Boeres, A.; Massacessi, L.; Istas, G.; Ventura, M.R.; Nunes Dos Santos, C.; Heiss, C.; Rodriguez-Mateos, A. Identification and quantification of novel cranberry-derived plasma and urinary (poly)phenols. *Arch. Biochem. Biophys.* **2016**, *599*, 31–41. [CrossRef] [PubMed]

69. Herszenyi, L.; Barabas, L.; Miheller, P.; Tulassay, Z. Colorectal cancer in patients with inflammatory bowel disease: The true impact of the risk. *Dig. Dis.* **2015**, *33*, 52–57. [CrossRef] [PubMed]

70. Pabla, B.; Bissonnette, M.; Konda, V.J. Colon cancer and the epidermal growth factor receptor: Current treatment paradigms, the importance of diet, and the role of chemoprevention. *World J. Clin. Oncol.* **2015**, *6*, 133–141. [CrossRef] [PubMed]

71. Duckworth, C.A.; Burkitt, M.D.; Williams, J.M.; Parsons, B.N.; Tang, J.M.; Pritchard, D.M. Murine models of Helicobacter (pylori or felis)-associated gastric cancer. *Curr. Protoc. Pharmacol.* **2015**, *69*, 14.34.1–14.34.35. [PubMed]

72. Iswaldi, I.; Arraez-Roman, D.; Gomez-Caravaca, A.M.; Contreras, M.D.; Uberos, J.; Segura-Carretero, A.; Fernandez-Gutierrez, A. Identification of polyphenols and their metabolites in human urine after cranberry-syrup consumption. *Food Chem. Toxicol.* **2013**, *55*, 484–492. [CrossRef] [PubMed]

73. Ohnishi, R.; Ito, H.; Kasajima, N.; Kaneda, M.; Kariyama, R.; Kumon, H.; Hatano, T.; Yoshida, T. Urinary excretion of anthocyanins in humans after cranberry juice ingestion. *Biosci. Biotechnol. Biochem.* **2006**, *70*, 1681–1687. [CrossRef] [PubMed]

74. Wang, C.J.; Zuo, Y.G.; Vinson, J.A.; Deng, Y.W. Absorption and excretion of cranberry-derived phenolics in humans. *Food Chem.* **2012**, *132*, 1420–1428. [CrossRef]

75. Zhang, K.; Zuo, Y.G. GC-MS determination of flavonoids and phenolic and benzoic acids in human plasma after consumption of cranberry juice. *J. Agric. Food Chem.* **2004**, *52*, 222–227. [CrossRef] [PubMed]

76. Blumberg, J.B.; Basu, A.; Krueger, C.G.; Lila, M.A.; Neto, C.C.; Novotny, J.A.; Reed, J.D.; Rodriguez-Mateos, A.; Toner, C.D. Impact of cranberries on gut microbiota and cardiometabolic health: Proceedings of the cranberry health research conference 2015. *Adv. Nutr.* **2016**, *7*, 759S–770S. [CrossRef]

77. The Cancer Genome Atlas Network. Comprehensive genomic characterization of head and neck squamous cell carcinomas. *Nature* **2015**, *517*, 576–582.

78. Feldman, M.; Tanabe, S.; Howell, A.; Grenier, D. Cranberry proanthocyanidins inhibit the adherence properties of Candida albicans and cytokine secretion by oral epithelial cells. *BMC Complement. Altern. Med.* **2012**, *12*, 6. [CrossRef] [PubMed]

79. Kresty, L.; Mallery, S.; Stoner, G. Black raspberries in cancer clinical trials: Past, present and future. *J. Berry Res.* **2016**, *6*, 251–261. [CrossRef]

80. Anhe, F.F.; Roy, D.; Pilon, G.; Dudonne, S.; Matamoros, S.; Varin, T.V.; Garofalo, C.; Moine, Q.; Desjardins, Y.; Levy, E.; et al. A polyphenol-rich cranberry extract protects from diet-induced obesity, insulin resistance and intestinal inflammation in association with increased Akkermansia spp. Population in the gut microbiota of mice. *Gut* **2015**, *64*, 872–883. [CrossRef] [PubMed]

antioxidants

MDPI

Article

Tart Cherry Extracts Reduce Inflammatory and Oxidative Stress Signaling in Microglial Cells

Barbara Shukitt-Hale *, Megan E. Kelly, Donna F. Bielinski and Derek R. Fisher

USDA-ARS, Human Nutrition Research Center on Aging, Tufts University, 711 Washington Street, Boston, MA 02111, USA; megan.kelly@ars.usda.gov (M.E.K.); donna.bielinski@tufts.edu (D.F.B.); derek.fisher@ars.usda.gov (D.R.F.)

* Correspondance: barbara.shukitthale@ars.usda.gov; Tel.: +1-617-556-3118

Academic Editor: Dorothy Klimis-Zacas
Received: 7 July 2016; Accepted: 9 September 2016; Published: 22 September 2016

Abstract: Tart cherries contain an array of polyphenols that can decrease inflammation and oxidative stress (OS), which contribute to cognitive declines seen in aging populations. Previous studies have shown that polyphenols from dark-colored fruits can reduce stress-mediated signaling in BV-2 mouse microglial cells, leading to decreases in nitric oxide (NO) production and inducible nitric oxide synthase (iNOS) expression. Thus, the present study sought to determine if tart cherries—which improved cognitive behavior in aged rats—would be efficacious in reducing inflammatory and OS signaling in HAPI rat microglial cells. Cells were pretreated with different concentrations (0–1.0 mg/mL) of Montmorency tart cherry powder for 1–4 h, then treated with 0 or 100 ng/mL lipopolysaccharide (LPS) overnight. LPS application increased extracellular levels of NO and tumor necrosis factor-alpha (TNF-α), and intracellular levels of iNOS and cyclooxygenase-2 (COX-2). Pretreatment with tart cherry decreased levels of NO, TNF-α, and COX-2 in a dose- and time-dependent manner versus those without pretreatment; the optimal combination was between 0.125 and 0.25 mg/mL tart cherry for 2 h. Higher concentrations of tart cherry powder and longer exposure times negatively affected cell viability. Therefore, tart cherries (like other dark-colored fruits), may be effective in reducing inflammatory and OS-mediated signals.

Keywords: antioxidant; anti-inflammatory; polyphenols; anthocyanins; cytokines

---

1. Introduction

Oxidative stress (OS) and inflammation in the brain contribute to the decline of motor abilities and cognitive performance with age [1,2]. While OS and inflammation both increase with age, the body's ability to defend and repair itself decreases through the lifespan [3,4]. This makes the brain more susceptible to the deleterious effects of OS and inflammation. Increased intake of antioxidants and anti-inflammatory compounds are believed to protect against this decline [5]. In this regard, certain foods with antioxidant and anti-inflammatory effects have been shown to help protect against the negative effects of aging.

Polyphenols are compounds from plants involved in antioxidant and anti-inflammatory cell activities that may be responsible (via an array of health-related bioactivities) for the multitude of beneficial effects that have been reported due to fruit and vegetable consumption [6]. There are thousands of different polyphenols found in plants, which are categorized into several groups based on their unique molecular structure. The presence of a number of bioactive compounds, including polyphenols, suggest tart cherries as a potential nutritional therapeutic to curtail the negative effects of aging. Tart cherries are rich in anthocyanins (one class of polyphenols), with cyanidin being the most abundant [7,8], as well as flavan-3-ols and flavonols [9]. The components of cherries may act directly to improve brain cell function and signaling, and/or may be more generally affecting extra-neuronal parameters of survival—such as inflammation—within the aging brain to improve behavior. In addition, cherries have been shown to reduce inflammation [10–13] and decrease oxidative stress [14–16].

Research with tart cherries and other dark-colored fruit has shown that their polyphenols become available to humans and rats in the bloodstream after consumption [17]. Consumption increases the levels of antioxidants and anti-inflammatory compounds due to the polyphenols from the tart cherries. Furthermore, it has been shown that cherry anthocyanins accumulate in the brain of young rats after 3 weeks of feeding with either 1% or 10% tart cherry-supplemented diets in a dose-dependent manner [18], and that dietary supplementation of tart cherries to rats aged 19–21 months improves age-related deficits in behavioral and neuronal functioning [19]. Similar age-related improvements were found when diets were supplemented with other dark-colored fruits, such as blueberry, strawberry, spinach, blackberries, cranberries, black currants, and Concord grape juice [20–24].

When lipopolysaccharide (LPS) is applied to microglial cells, it activates the cells in the same way as what occurs during a bacterial infection, which increases OS and inflammation. Increased inflammation leads to increased levels of inflammatory stress signals, such as nitric oxide (NO), tumor necrosis factor-alpha (TNF-α), and cyclooxygenase-2 (COX-2), as well as increased expression of inducible nitric oxide synthase (iNOS). A previous study using these markers showed the anti-inflammatory and antioxidant effects of pretreatment of BV-2 mouse microglial cells with açai pulp fractions, which are also rich in anthocyanins [25]. After açai pretreatment, the cells were treated with LPS. These cells showed reduced production of NO, TNF-α, and COX-2, and reduced iNOS expression relative to cells not pretreated with açai extracts, showing that dark-colored fruits rich in polyphenols can protect against OS and inflammation in microglial cells.

The present study sought to determine whether the application of tart cherry to HAPI rat microglial cells would protect against LPS-induced OS and inflammation in a similar fashion. To more closely relate to our behavioral study with tart cherry [19], we used HAPI cells in these experiments because they are derived from rats, as opposed to the BV-2 cells, which are derived from mice. HAPI rat microglial cells were pretreated for various durations and concentrations with tart cherry powder. After subsequent treatment with LPS, OS and inflammation were determined by measuring levels of NO, TNF-α, COX-2, and iNOS. Furthermore, this study sought to determine the ideal tart cherry concentration and pretreatment time to minimize OS and inflammation.

2. Materials and Methods

2.1. Cell Culture

HAPI rat microglial cells (generously provided by Dr. Grace Sun, University of Missouri, Columbia, MO, USA) were maintained in Dulbecco's modified Eagle's medium (DMEM, Invitrogen, Grand Island, NY, USA) supplemented with 10% fetal bovine serum (FBS), 100 U/mL penicillin, and 100 ug/mL streptomycin at 37 °C in a humidified incubator under 5% CO_2. Cells were grown in 100 mm plates and then split into 12-well plates prior to treatment. All treatment groups were assayed in duplicate. For experiments, cells were incubated in DMEM without phenol red. Cells were then pretreated with Montmorency tart cherry powder (0.062, 0.125, 0.250, 0.500, 1.000 mg/mL) diluted in media, or with control media for 1, 2, or 4 h. After the pretreatment, the media was removed and cells were stimulated overnight with lipopolysaccharide (LPS, Sigma-Aldrich, St. Louis, MO, USA) at 100 ng/mL or 0 ng/mL in DMEM without phenol red. The freeze-dried Montmorency tart cherry powder (*Prunus cerasus* L.) was provided by the Cherry Marketing Institute (Dewitt, MI, USA), and its polyphenolic composition has been previously published [9,26]. Additionally, HAPI cells have been used in previous studies to study the effects of LPS-induced inflammation and possible neuroprotection [27–29].

2.2. Viability

Cell viability was measured using Live/Dead Cellular Viability/Cytotoxicity Kit (Molecular Probes, Eugene, OR, USA). Calcein AM labels live cells with intact membranes with a green fluorescent compound. Ethidium homodimer-1 labels dead cells with damaged membranes with red fluorescence. Fluorescent images of the cell were captured with a Nikon TE2000U inverted fluorescent microscope.

2.3. Nitrite Quantification

Under physiological conditions, NO is oxidized into nitrite (NO_2^-). Therefore, to assess the production of NO from HAPI cells after the pretreatments and treatments described above, the extracellular release of nitrite was measured by Greiss reagent (Invitrogen) according to manufacturer's instructions. Absorbance was read at 548 nm, and the concentration of nitrite was calculated with the linear equation derived from the standard curve generated by known concentrations of nitrite.

2.4. Western Blots

Cells were washed in ice-cold PBS, re-suspended, and lysed by agitation in CelLytic-M Cell Lysis Reagent (Sigma). Cells were then centrifuged at $18,000 \times g$ for 10 min at 4 °C. The resultant supernatant lysate was used for blotting, and the pellet was discarded. Western blots were performed as described by Poulose and colleagues [25], except that 10% polyacrylamide gels were used. Primary antibodies for iNOS (Millipore, Billerica, MA, USA) and COX-2 (Santa Cruz, Dallas, TX, USA) were used at 1:1000 dilutions for incubation overnight at 4 °C. Following enhanced chemiluminescence (ECL) development, the optical density of antibody-specific bands was analyzed by the LabWorks image acquisition and analysis software (UVP, Upland, CA, USA).

2.5. TNF-α ELISA

Quantification of tumor necrosis factor-alpha (TNF-α) in cell-conditioned media was performed with an enzyme-linked immunosorbent assay (ELISA, eBioscience, San Diego, CA, USA) according to the manufacturer's instructions. TNF-α concentration for each sample was calculated from the linear equation derived from the standard curve of known concentrations of the cytokine.

2.6. Data Analysis

All statistical analyses were performed using SYSTAT software (SPSS, Inc., Chicago, IL, USA). Data are expressed as mean ± SEM. The data were analyzed by three way analyses of variance (ANOVA) with tart cherry dose, duration, and LPS exposure as experimental factors, followed by post hoc testing with Tukey's HSD test to determine differences among groups. Results were considered statistically significant if the observed significance level *p* was <0.05. Note that for each dependent measure, those cells treated with LPS alone were statistically higher than the control conditions without LPS, which were not different (data not shown for the no LPS condition). Additionally, pretreatment with tart cherry did not significantly affect cells in the absence of LPS in any of the endpoints assayed (data not shown).

3. Results

3.1. Viability

Higher concentrations of tart cherry powder and longer exposure times led to increased cell death. The images (see Appendix A, Figure A1) showed that 1.000 mg/mL of tart cherry powder was too high of a concentration for treatment, and therefore only the lower concentrations of 0.062, 0.125, 0.250, and 0.500 mg/mL were used for testing protection against LPS-induced inflammation and OS. The 0.500 mg/mL dose at 2 and 4 h also caused minor decreased viability. There was no change in viability in control cells (a dose of 0 cherry and 0 LPS) or cells treated with 100 ng/mL LPS, showing that this dose of LPS did not result in cell death, as has been shown previously [29].

3.2. Nitric Oxide

Results showed that tart cherry pretreatment attenuated LPS-induced nitric oxide (NO) production in HAPI microglia in a dose- and time-dependent manner (Figure 1). NO is a free radical that can also act as a second messenger involved in a number of functions, including cellular immune response and the activation of apoptosis. LPS application significantly increased nitrite release in all groups ($p < 0.05$ compared to control conditions without LPS). However, LPS-induced nitrite release was significantly ($p < 0.05$) reduced by pretreatment with doses of 0.250 and 0.500 mg/mL cherry at 1 h, and doses of 0.125, 0.250, and 0.500 mg/mL cherry at 2 and 4 h, compared to LPS alone. This decrease in NO release at doses of 0.500 mg/mL at 2 and 4 h, however, is most likely due to the decreased viability at this concentration and time, as shown previously.

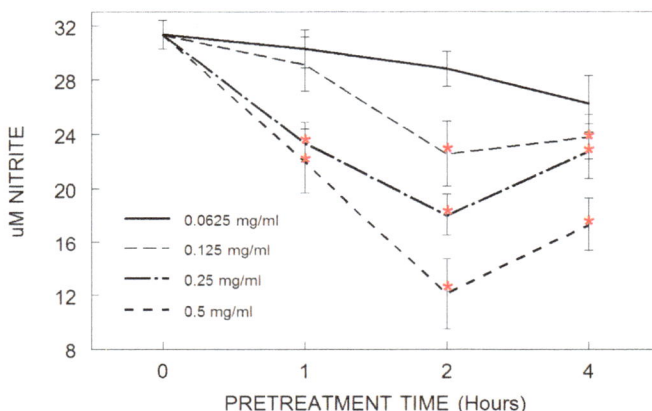

Figure 1. Production of extracellular nitric oxide (NO) in HAPI microglial cells pretreated with tart cherry powder (0.062, 0.125, 0.250, or 0.500 mg/mL) for 0 (LPS only/alone control), 1, 2, or 4 h, and stimulated overnight with lipopolysaccharide (LPS, 100 ng/mL). Data are expressed as mean ± SEM and quantified using the Greiss reagent. Each tart cherry pretreatment was compared against LPS treatment alone. Comparative post hoc analyses were made by Tukey's HSD with significance at (*) $p < 0.05$ versus LPS.

3.3. iNOS

Inducible nitric oxide synthase (iNOS) produces the inflammatory mediator nitric oxide (NO). LPS application increased iNOS expression in all groups compared to control conditions not exposed to LPS ($p < 0.05$). However, tart cherry pretreatment did not protect against these LPS-induced increases at any concentration or exposure time (Figure 2).

3.4. TNF-α

Tart cherry pretreatment reduced the LPS-induced release of the inflammatory cytokine tumor necrosis factor-alpha (TNF-α) in HAPI microglia in a dose- and time-dependent manner (Figure 3). LPS application increased TNF-α release in all groups compared to control groups with no LPS ($p < 0.05$). This increase was significantly reduced by pretreatment with concentrations of 0.062 and 0.250 mg/mL of tart cherry for 1 h, and all doses of cherry at 2 and 4 h, compared to LPS alone. At 4 h, however, TNF-α levels began to rise relative to the 2 h pretreatment, suggesting that 4 h was too long a duration for pretreatment.

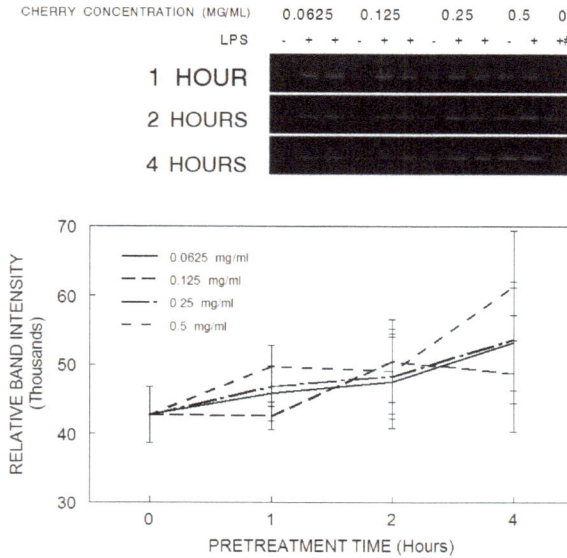

Figure 2. Expression of inducible nitric oxide synthase (iNOS) in HAPI microglial cells pretreated with tart cherry powder (0.062, 0.125, 0.250, or 0.500 mg/mL) for 0 (LPS only/alone control), 1, 2, or 4 h and stimulated overnight with lipopolysaccharide (LPS, 100 ng/mL). Quantitative measurements were made based on the Western blots, and are expressed as mean ± SEM for the immunoreactive band density. Each tart cherry pretreatment was compared against LPS treatment alone, and comparative post hoc analyses were made by Tukey's HSD. [#] column represents blots for 0 cherry with either no LPS (in the 1 h row) or LPS (in the 2 and 4 h row).

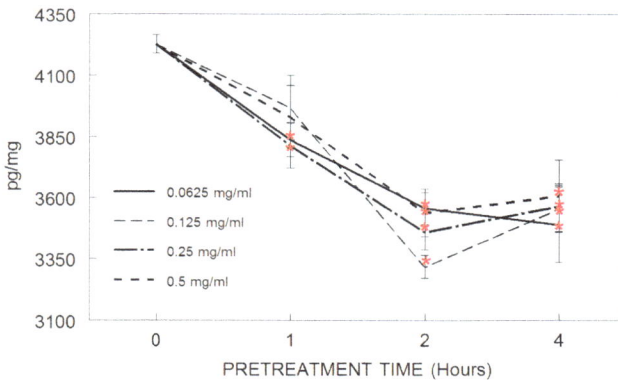

Figure 3. Suppression of tumor necrosis factor-alpha (TNF-α) release in HAPI microglial cells pretreated with tart cherry powder (0.062, 0.125, 0.250, or 0.500 mg/mL) for 0 (LPS only/alone control), 1, 2, or 4 h and stimulated overnight with lipopolysaccharide (LPS, 100 ng/mL). Data are expressed as mean ± SEM (pg/mg of media) as assayed by enzyme-linked immunosorbent assay (ELISA). Each tart cherry pretreatment was compared against LPS treatment alone. Comparative post hoc analyses were made by Tukey's HSD with significance at (*) $p < 0.05$ versus LPS.

3.5. COX-2

Results showed that lower doses of tart cherry pretreatment reduced LPS-induced cyclooxygenase-2 (COX-2) expression in HAPI microglia at shorter durations (Figure 4). COX-2 is responsible for the formation of prostanoids, which are inflammatory mediators. LPS application increased COX-2 expression in all groups compared to control conditions not exposed to LPS ($p < 0.05$). This increase was significantly reduced with doses of 0.062 and 0.125 mg/mL cherry at 2 h, compared to LPS alone. Conversely, pretreatment with 0.5 mg/mL cherry for 1 h significantly increased COX-2 expression.

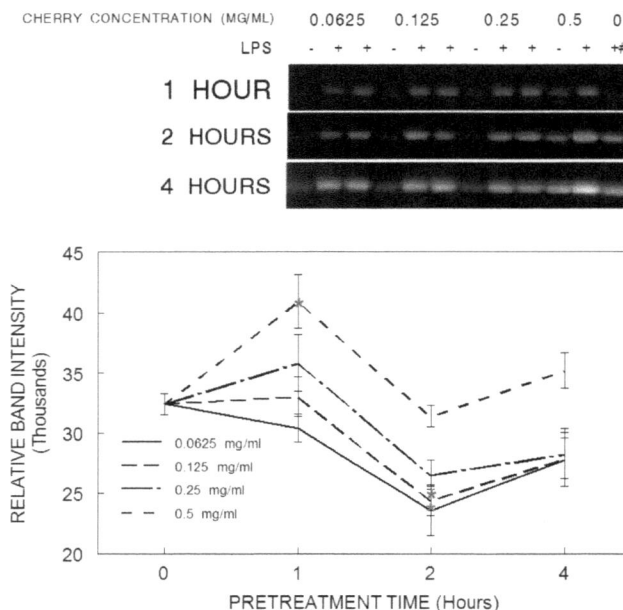

Figure 4. Reduction in cyclooxygenase-2 (COX-2) expression in HAPI microglial cells pretreated with tart cherry powder (0.062, 0.125, 0.250, or 0.500 mg/mL) for 0 (LPS only/alone control), 1, 2, or 4 h and stimulated overnight with lipopolysaccharide (LPS, 100 ng/mL). Quantitative measurements were made based on the Western blots, and are expressed as mean ± SEM for the immunoreactive band density. Each tart cherry pretreatment was compared against LPS treatment alone. Comparative post hoc analyses were made by Tukey's HSD with significance at (*) $p < 0.05$ versus LPS. # column represents blots for 0 cherry with either no LPS (in the 1 h row) or LPS (in the 2 and 4 h row).

4. Discussion

The results of this study show that pretreating HAPI microglia cells with Montmorency tart cherry powder enhanced protection against oxidative and inflammatory stress by reducing stress-mediated signaling in a dose- and time-dependent manner. Because higher doses and longer treatment durations negatively affected cell viability and some stress-mediated signals, a concentration of between 0.125 and 0.25 mg/mL tart cherry, at a treatment time of 2 h, appeared to be the optimal combination.

Microglia mediate inflammation as a response to stress or injury in the central nervous system. This response can be beneficial, such as by recruiting bone marrow-derived cells to an injured brain region to aid in the healing process [30]. Alternatively, microglial activation can be problematic, due to the release of potentially neurotoxic substances [31]. Furthermore, inflammation in the central nervous system has been linked to autoimmune diseases such as multiple sclerosis [32] and central nervous system degenerative diseases [33], as well as accelerating disease progression and worsening symptoms [3]. Microglia, when in a highly activated state—such as that caused by stressors like LPS—produce inflammatory molecules such as cytokines, superoxide, and nitric oxide, ultimately leading to a cascade of pro-inflammatory proteins and cell death [34].

LPS application activates microglial cells, leading to the release of NO and TNF-α, which together can be neurotoxic [31]. NO is an inflammatory mediator released by iNOS. TNF-α is a cytokine which also mediates inflammation. Similarly, COX-2 is an enzyme that catalyzes the reaction that creates prostanoids, which in turn increase cellular inflammation [35]. LPS application to microglial cells increases the release of all four of these inflammatory mediators compared to cells not treated with LPS. However, significantly less NO was released from cells pretreated with doses of tart cherry powder ranging from 0.125 to 0.500 mg/mL for 1–4 h. The lower NO release at 0.500 mg/mL for 2 and 4 h, however, is likely due to decreased cell viability at these times and this concentration, because there are fewer cells that are present and therefore able to release NO. The shorter durations and lower concentrations of tart cherry did not cause cell death, but still lowered extracellular NO levels compared to controls not pretreated with cherry. Similarly, COX-2 expression was significantly reduced when pretreated with the lowest tart cherry doses for 2 h compared to cells treated with high doses of tart cherry powder or not treated at all. Extracellular TNF-α levels were also reduced for 0.062 and 0.125 mg/mL cherry at 1 h and at all concentrations at 2 h and 4 h. The 4-h pretreatment, however, was again too long a duration to pretreat the cells, and the lower TNF-α levels are most likely due to the decreased number of viable cells. The application of cherry powder did not significantly affect iNOS production in LPS-exposed cells. This result was unexpected, because iNOS produces NO, and NO levels were lowered by tart cherry application. The reason behind the reduction in NO production without concomitant reduction in iNOS is not clear, but one possibility is that tart cherry polyphenols could be affecting the activity of iNOS or the levels of its cofactors, rather than iNOS expression. Taken together, these results suggest that pretreatment with Montmorency tart cherry powder helps to reduce inflammation and OS in HAPI rat microglial cells after exposure to a known inflammatory substance in a dose- and time-dependent manner. Similar neuroprotective results have been seen with other food extracts, such as walnuts [36,37], blueberries [38], and açai [25].

Exposure to conditions that produce oxidative and inflammatory stress causes symptoms that mimic those traditionally seen in aging populations [39,40]. As humans age, our ability to defend against these substances and their effects weakens, putting elderly people at increased risk for neuronal disease and degradation [3]. Indeed, substantial deficits in cognitive and motor performance have been shown in older populations [41]. Therefore, oxidative stress and inflammation should be minimized in the brain to avoid these possible negative outcomes. One way to protect an aging brain against this damage is to curtail microglial activation with neuroprotective foods, such as tart cherries. Polyphenols are one class of food-derived chemicals that may protect microglia against detrimental activation [5,25]. In our model, the application of tart cherry powder to HAPI rat microglial cells provided access to the polyphenols from the cherries, which protected the cells against the deleterious effects of LPS, a known inducer of inflammation in microglia. Therefore, tart cherries (like other dark-colored fruits) may be effective in reducing inflammation and oxidative stress in the brain, thereby protecting against cognitive declines in aged populations. This protection might be one mechanism by which dietary supplementation of tart cherries to rats aged 19–21 months improved age-related deficits in behavioral and neuronal functioning [19]. Because of the known age-related consequences of inflammation and oxidative stress, it may be important to consume neuroprotective foods—such as those rich in polyphenols—to deter cognitive decline. These results, coupled with our previous studies, show that the addition of cherries to the diet may increase "health span" in aging, and may slow the aging process by reducing the incidence and/or delaying the onset of debilitating neurodegenerative disease.

5. Conclusions

In conclusion, Montmorency tart cherry powder, which is high in polyphenols, was effective in reducing inflammatory and oxidative stress signaling in HAPI rat microglial cells. Pretreatment with tart cherry decreased levels of LPS-induced NO, TNF-α, and COX-2 in a dose- and time-dependent manner versus those without pretreatment; the optimal combination was between 0.125 and 0.25 mg/mL tart cherry for 2 h. Therefore, tart cherries, like other dark-colored fruits, may be effective in reducing inflammatory and OS-mediated signals.

Acknowledgments: This research was supported by USDA Intramural funds and an agreement between the USDA and the Cherry Marketing Institute.

Author Contributions: Barbara Shukitt-Hale and Derek R. Fisher conceived and designed the experiments; Megan E. Kelly, Donna F. Bielinski and Derek R. Fisher performed the experiments; Derek R. Fisher analyzed the data; Barbara Shukitt-Hale and Megan E. Kelly wrote the paper.

Conflicts of Interest: The authors declare no conflict of interest. The funding sponsors had no role in the design of the study; in the collection, analyses, or interpretation of data; in the writing of the manuscript, and in the decision to publish the results.

Appendix A

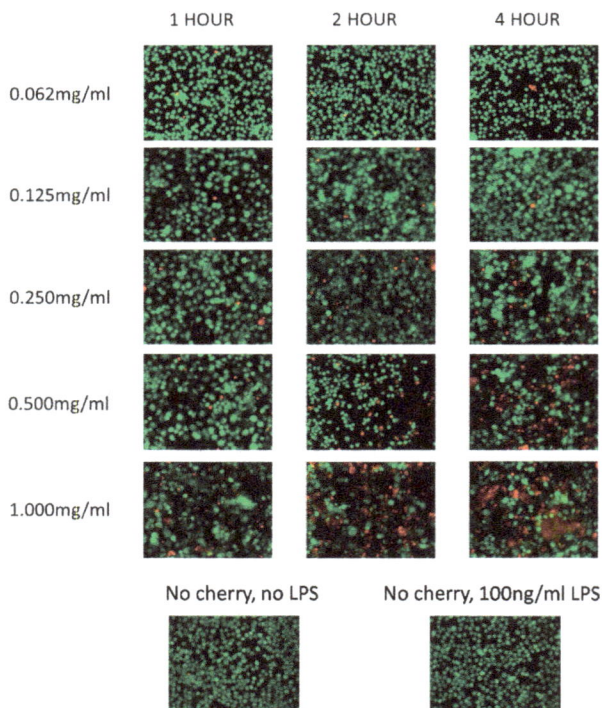

Figure A1. Effect of tart cherry powder on cellular viability. Representative images of cells treated with tart cherry powder (0.062, 0.125, 0.250, 0.500, or 1.000 mg/mL) for 1, 2, or 4 h, as well as cells treated with 0 cherry and 0 or 100 ng/mL LPS. Live cells with intact membranes are labeled green with Calcein AM, while dead cells with damaged membranes are labeled red with ethidium homodimer-1. Images were taken under a fluorescent microscope at 80× magnification.

References

1. Hauss-Wegrzyniak, B.; Vannucchi, M.G.; Wenk, G.L. Behavioral and ultrastructural changes induced by chronic neuroinflammation in young rats. *Brain Res.* **2000**, *859*, 157–166. [CrossRef]
2. Campbell, N.R.; Burgess, E.; Choi, B.C.; Taylor, G.; Wilson, E.; Cleroux, J.; Fodor, J.G.; Leiter, L.A.; Spence, D. Lifestyle modifications to prevent and control hypertension. 1. Methods and an overview of the Canadian recommendations. Canadian hypertension society, Canadian coalition for high blood pressure prevention and control, laboratory centre for disease control at health Canada, heart and stroke foundation of Canada. *CMAJ* **1999**, *160*, S1–S6. [PubMed]
3. Perry, V.H. Contribution of systemic inflammation to chronic neurodegeneration. *Acta Neuropathol.* **2010**, *120*, 277–286. [CrossRef] [PubMed]
4. Romano, A.D.; Serviddio, G.; de Matthaeis, A.; Bellanti, F.; Vendemiale, G. Oxidative stress and aging. *J. Nephrol.* **2010**, *23*, S29–S36. [PubMed]
5. Joseph, J.A.; Shukitt-Hale, B.; Casadesus, G. Reversing the deleterious effects of aging on neuronal communication and behavior: Beneficial properties of fruit polyphenolic compounds. *Am. J. Clin. Nutr.* **2005**, *81*, 313S–316S. [PubMed]
6. Stevenson, D.E.; Hurst, R.D. Polyphenolic phytochemicals—Just antioxidants or much more? *Cell. Mol. Life Sci.* **2007**, *64*, 2900–2916. [CrossRef] [PubMed]
7. Manach, C.; Scalbert, A.; Morand, C.; Remesy, C.; Jimenez, L. Polyphenols: Food sources and bioavailability. *Am. J. Clin. Nutr.* **2004**, *79*, 727–747. [PubMed]
8. Seeram, N.P.; Bourquin, L.D.; Nair, M.G. Degradation products of cyanidin glycosides from tart cherries and their bioactivities. *J. Agric. Food Chem.* **2001**, *49*, 4924–4929. [CrossRef] [PubMed]
9. Bhagwat, S.; Haytowitz, D.; Holden, J. *USDA Database for the Flavonoid Content of Selected Foods Release 3.1*; Food and Agriculture Organization of the United Nations: Rome, Italy, 2014.
10. Seeram, N.P.; Momin, R.A.; Nair, M.G.; Bourquin, L.D. Cyclooxygenase inhibitory and antioxidant cyanidin glycosides in cherries and berries. *Phytomedicine* **2001**, *8*, 362–369. [CrossRef] [PubMed]
11. Tall, J.M.; Seeram, N.P.; Zhao, C.; Nair, M.G.; Meyer, R.A.; Raja, S.N. Tart cherry anthocyanins suppress inflammation-induced pain behavior in rat. *Behav. Brain Res.* **2004**, *153*, 181–188. [CrossRef] [PubMed]
12. Ou, B.; Bosak, K.N.; Brickner, P.R.; Iezzoni, D.G.; Seymour, E.M. Processed tart cherry products—Comparative phytochemical content, in vitro antioxidant capacity and in vitro anti-inflammatory activity. *J. Food Sci.* **2012**, *77*, H105–H112. [CrossRef] [PubMed]
13. Seymour, E.M.; Lewis, S.K.; Urcuyo-Llanes, D.E.; Tanone, I.I.; Kirakosyan, A.; Kaufman, P.B.; Bolling, S.F. Regular tart cherry intake alters abdominal adiposity, adipose gene transcription, and inflammation in obesity-prone rats fed a high fat diet. *J. Med. Food* **2009**, *12*, 935–942. [CrossRef] [PubMed]
14. Kim, D.O.; Heo, H.J.; Kim, Y.J.; Yang, H.S.; Lee, C.Y. Sweet and sour cherry phenolics and their protective effects on neuronal cells. *J. Agric. Food Chem.* **2005**, *53*, 9921–9927. [CrossRef] [PubMed]
15. Traustadottir, T.; Davies, S.S.; Stock, A.A.; Su, Y.; Heward, C.B.; Roberts, L.J., II; Harman, S.M. Tart cherry juice decreases oxidative stress in healthy older men and women. *J. Nutr.* **2009**, *139*, 1896–1900. [CrossRef] [PubMed]
16. Wang, H.; Nair, M.G.; Strasburg, G.M.; Chang, Y.C.; Booren, A.M.; Gray, J.I.; DeWitt, D.L. Antioxidant and antiinflammatory activities of anthocyanins and their aglycon, cyanidin, from tart cherries. *J. Nat. Prod.* **1999**, *62*, 294–296. [CrossRef] [PubMed]
17. Miyazawa, T.; Nakagawa, K.; Kudo, M.; Muraishi, K.; Someya, K. Direct intestinal absorption of red fruit anthocyanins, Cyanidin-3-glucoside and Cyanidin-3,5-diglucoside, into rats and humans. *J. Agric. Food. Chem.* **1999**, *47*, 1083–1091. [CrossRef] [PubMed]
18. Kirakosyan, A.; Seymour, E.M.; Wolforth, J.; McNish, R.; Kaufman, P.B.; Bolling, S.F. Tissue bioavailability of anthocyanins from whole tart cherry in healthy rats. *Food Chem.* **2015**, *171*, 26–31. [CrossRef] [PubMed]
19. Thangthaeng, N.; Poulose, S.M.; Gomes, S.M.; Miller, M.G.; Bielinski, D.F.; Shukitt-Hale, B. Tart cherry supplementation improves working memory, hippocampal inflammation, and autophagy in aged rats. *AGE* **2016**. [CrossRef] [PubMed]
20. Joseph, J.A.; Shukitt-Hale, B.; Denisova, N.A.; Bielinski, D.; Martin, A.; McEwen, J.J.; Bickford, P.C. Reversals of age-related declines in neuronal signal transduction, cognitive, and motor behavioral deficits with blueberry, spinach, or strawberry dietary supplementation. *J. Neurosci.* **1999**, *19*, 8114–8121. [PubMed]

21. Shukitt-Hale, B.; Cheng, V.; Joseph, J.A. Effects of blackberries on motor and cognitive function in aged rats. *Nutr. Neurosci.* **2009**, *12*, 135–140. [CrossRef] [PubMed]

22. Shukitt-Hale, B.; Galli, R.; Meterko, V.; Carey, A.; Bielinski, D.; McGhie, T.; Joseph, J.A. Dietary supplementation with fruit polyphenolics ameliorates age-related deficits in behavior and neuronal markers of inflammation and oxidative stress. *AGE* **2005**, *27*, 49–57. [CrossRef] [PubMed]

23. Shukitt-Hale, B.; Carey, A.; Simon, L.; Mark, D.A.; Joseph, J.A. Effects of concord grape juice on cognitive and motor deficits in aging. *Nutrition* **2006**, *22*, 295–302. [CrossRef] [PubMed]

24. Shukitt-Hale, B.; Bielinski, D.F.; Lau, F.C.; Willis, L.M.; Carey, A.N.; Joseph, J.A. The beneficial effects of berries on cognition, motor behaviour and neuronal function in ageing. *Br. J. Nutr.* **2015**, *114*, 1542–1549. [CrossRef] [PubMed]

25. Poulose, S.M.; Fisher, D.R.; Larson, J.; Bielinski, D.F.; Rimando, A.M.; Carey, A.N.; Schauss, A.G.; Shukitt-Hale, B. Anthocyanin-rich Açai (*Euterpe oleracea* Mart.) fruit pulp fractions attenuate inflammatory stress signaling in mouse brain BV-2 microglial cells. *J. Agric. Food Chem.* **2012**, *60*, 1084–1093. [CrossRef] [PubMed]

26. Mulabagal, V.; Lang, G.A.; DeWitt, D.L.; Dalavoy, S.S.; Nair, M.G. Anthocyanin content, lipid peroxidation and cyclooxygenase enzyme inhibitory activities of sweet and sour cherries. *J. Agric. Food Chem.* **2009**, *57*, 1239–1246. [CrossRef] [PubMed]

27. Wang, C.; Nie, X.; Zhang, Y.; Li, T.; Mao, J.; Liu, X.; Gu, Y.; Shi, J.; Xiao, J.; Wan, C.; et al. Reactive oxygen species mediate nitric oxide production through ERK/JNK MAPK signaling in hapi microglia after PFOS exposure. *Toxicol. Appl. Pharmacol.* **2015**, *288*, 143–151. [CrossRef] [PubMed]

28. Zheng, W.; Zheng, X.; Liu, S.; Ouyang, H.; Levitt, R.C.; Candiotti, K.A.; Hao, S. Tnfalpha and IL-1beta are mediated by both TLR4 and Nod1 pathways in the cultured HAPI cells stimulated by LPS. *Biochem. Biophys. Res. Commun.* **2012**, *420*, 762–767. [CrossRef] [PubMed]

29. Jantaratnotai, N.; Utaisincharoen, P.; Sanvarinda, P.; Thampithak, A.; Sanvarinda, Y. Phytoestrogens mediated anti-inflammatory effect through suppression of IRF-1 and PSTAT1 expressions in lipopolysaccharide-activated microglia. *Int. Immunopharmacol.* **2013**, *17*, 483–488. [CrossRef] [PubMed]

30. Ladeby, R.; Wirenfeldt, M.; Garcia-Ovejero, D.; Fenger, C.; Dissing-Olesen, L.; Dalmau, I.; Finsen, B. Microglial cell population dynamics in the injured adult central nervous system. *Brain Res. Rev.* **2005**, *48*, 196–206. [CrossRef] [PubMed]

31. Hemmer, K.; Fransen, L.; Vanderstichele, H.; Vanmechelen, E.; Heuschling, P. An in vitro model for the study of microglia-induced neurodegeneration: Involvement of nitric oxide and tumor necrosis factor-alpha. *Neurochem. Int.* **2001**, *38*, 557–565. [CrossRef]

32. Gschwandtner, M.; Piccinini, A.M.; Gerlza, T.; Adage, T.; Kungl, A.J. Interfering with the CCL2-glycosaminoglycan axis as a potential approach to modulate neuroinflammation. *Neurosci. Lett.* **2016**, *626*, 164–173. [CrossRef] [PubMed]

33. Feng, Y.; Xue, H.; Zhu, J.; Yang, L.; Zhang, F.; Qian, R.; Lin, W.; Wang, Y. ESE1 is associated with neuronal apoptosis in lipopolysaccharide induced neuroinflammation. *Neurochem. Res.* **2016**. [CrossRef] [PubMed]

34. Qin, L.; Wu, X.; Block, M.L.; Liu, Y.; Breese, G.R.; Hong, J.S.; Knapp, D.J.; Crews, F.T. Systemic LPS causes chronic neuroinflammation and progressive neurodegeneration. *Glia* **2007**, *55*, 453–462. [CrossRef] [PubMed]

35. O'Banion, M.K. Cyclooxygenase-2: Molecular biology, pharmacology, and neurobiology. *Crit. Rev. Neurobiol.* **1999**, *13*, 45–82. [PubMed]

36. Willis, L.M.; Bielinski, D.F.; Fisher, D.R.; Matthan, N.R.; Joseph, J.A. Walnut extract inhibits LPS-induced activation of BV-2 microglia via internalization of TLR4: Possible involvement of phospholipase D2. *Inflammation* **2010**, *33*, 325–333. [CrossRef] [PubMed]

37. Carey, A.N.; Fisher, D.R.; Joseph, J.A.; Shukitt-Hale, B. The ability of walnut extract and fatty acids to protect against the deleterious effects of oxidative stress and inflammation in hippocampal cells. *Nutr. Neurosci.* **2013**, *16*, 13–20. [CrossRef] [PubMed]

38. Joseph, J.A.; Shukitt-Hale, B.; Brewer, G.J.; Weikel, K.A.; Kalt, W.; Fisher, D.R. Differential protection among fractionated blueberry polyphenolic families against DA-, $A\beta_{42}$- and LPS-induced decrements in Ca^{2+} buffering in primary hippocampal cells. *J. Agric. Food Chem.* **2010**, *58*, 8196–8204. [CrossRef] [PubMed]

39. Shukitt-Hale, B.; Casadesus, G.; McEwen, J.J.; Rabin, B.M.; Joseph, J.A. Spatial learning and memory deficits induced by exposure to iron-56-particle radiation. *Radiat. Res.* **2000**, *154*, 28–33. [CrossRef]

40. Shukitt-Hale, B. The effects of aging and oxidative stress on psychomotor and cognitive behavior. *AGE* **1999**, *22*, 9–17. [CrossRef] [PubMed]

41. Miller, M.G.; Hamilton, D.A.; Joseph, J.A.; Shukitt-Hale, B. Mobility and cognition: End points for dietary interventions in aging. *Nutr. Aging* **2014**, *2*, 213–222.

antioxidants

MDPI

Article

Chemical Analysis of Extracts from Newfoundland Berries and Potential Neuroprotective Effects

Mohammad Z. Hossain, Emily Shea, Mohsen Daneshtalab † and John T. Weber *

School of Pharmacy, Memorial University of Newfoundland, 300 Prince Philip Drive, St. John's,
NL A1B 3V6, Canada; zahidrobin@yahoo.com (M.Z.H.); z25epms@mun.ca (E.S.); mohsen@mun.ca (M.D.)
* Correspondence: jweber@mun.ca; Tel.: +1-709-777-7022; Fax: +1-709-777-7044
† We sadly need to note here that Mohsen Daneshtalab passed away on 16 May 2014.

Academic Editor: Dorothy Klimis-Zacas
Received: 15 July 2016; Accepted: 8 October 2016; Published: 19 October 2016

Abstract: Various species of berries have been reported to contain several polyphenolic compounds, such as anthocyanins and flavonols, which are known to possess high antioxidant activity and may be beneficial for human health. To our knowledge, a thorough chemical analysis of polyphenolics in species of these plants native to Newfoundland, Canada has not been conducted. The primary objective of this study was to determine the polyphenolic compounds present in commercial extracts from Newfoundland berries, which included blueberries (*V. angustifolium*), lingonberries (*V. vitis-idaea*) and black currant (*Ribes lacustre*). Anthocyanin and flavonol glycosides in powdered extracts from *Ribes lacustre* and the *Vaccinium* species were identified using the high performance liquid chromatographic (HPLC) separation method with mass spectrometric (MS) detection. The identified compounds were extracted from dried berries by various solvents via ultrasonication followed by centrifugation. A reverse-phase analytical column was employed to identify the retention time of each chemical component before submission for LC–MS analysis. A total of 21 phenolic compounds were tentatively identified in the three species. Further, we tested the effects of the lingonberry extract for its ability to protect neurons and glia from trauma utilizing an in vitro model of cell injury. Surprisingly, these extracts provided complete protection from cell death in this model. These findings indicate the presence of a wide variety of anthocyanins and flavonols in berries that grow natively in Newfoundland. These powdered extracts maintain these compounds intact despite being processed from berry fruit, indicating their potential use as dietary supplements. In addition, these recent findings and previous data from our lab demonstrate the ability of compounds in berries to protect the nervous system from traumatic insults.

Keywords: anthocyanins; antioxidants; flavonols; *Ribes lacustre*; trauma; *Vaccinium* species

1. Introduction

Natural polyphenolic antioxidants have attracted the attention of food scientists because of their positive effects on human health. Many species of berries have high amounts of polyphenolic antioxidants and therefore, are important sources of potential health promoting components [1]. Flavonoids in particular are polyphenols that constitute a large group of secondary plant metabolites [2]. The predominant flavonoids found in berries are anthocyanins and flavonols, which are almost exclusively present in glycosylated forms [3]. The main anthocyanins in fruits are glycosides of six anthocyanidins, with cyanidin as the predominant anthocyanidin, followed by delphinidin, peonidin, pelargonidin, petunidin and malvidin [4–6]. Delphinidin is known to be responsible for bluish colours, whereas cyanidin and pelargonidin are responsible for red and purple colours, respectively, in the fruits and vegetables. These compounds have a wide range of biological effects, including potent antioxidant properties, which can protect cells against free radical attack [7,8]. The antioxidant capacity of flavonoids is influenced by the type of sugar moiety, degree of glycosylation, and acylation of anthocyanin glucosides [9,10]. Owing to the large number of flavonoid glycosides present in fruits and the lack of analytical standards available, acid hydrolysis has been used to cleave glycosidic bonds, followed by High Performance Liquid Chromatography (HPLC) in order to quantify the aglycones. A limitation associated with acid hydrolysis of flavonoid glycosides is that the acid concentrations, incubation time and temperature need to be optimised for different classes of flavonoids [11,12]. At present, the most satisfactory method for analysis of mixtures is the multi-step method of separation, isolation and quantification by Liquid Chromatography (LC) with peak identification by Mass spectrophotometry (MS) [3,13].

Detailed HPLC analysis has been conducted on several varieties of berries in various areas of Canada [14–16], the United States [3,17,18], and Finland [19–23]. The primary objective of this study was to determine the different polyphenolic components present in commercial extracts from *Ribes lacustre* and *Vaccinium* species that grow natively in Newfoundland, with a particular interest in compounds reported to have high antioxidant activity. We also tested the effects of the lingonberry extract for its ability to protect neurons and glia from trauma in vitro.

2. Materials and Methods

2.1. Samples and Chemicals

Dried powders of *Vaccinium angustifolium* (lowbush blueberry), *Ribes lacustre* (bristly black currant) and *Vaccinium vitis-idaea* (lingonberry; also known as partridgeberry in Newfoundland) were provided from Natural Newfoundland Nutraceuticals (NNN Inc., Markland, NL, Canada). These powders were produced using a Refractance Window® (MCD Technologies, Tacoma, WA, USA) drying technique [24]. The standard cyanidin 3-galactoside was purchased from Chromadex Inc. (Irvine, CA, USA). HPLC grade methanol and acetone were obtained from Caledon laboratories (Georgetown, ON, Canada). Formic acid (HPLC grade) was acquired from Fisher scientific (Fairlawn, NJ, USA).

2.2. Extract Preparation

Extracts were prepared similar to the methods of Cho et al. [3]. In brief, samples of berry powders (5 g) were first treated with 20 mL of extract-ion solution containing methanol/water/formic acid (60:37:3 $v/v/v$). Sonication was performed for 30 min using a Fisher Inc. sonicator (model FS110H) and the mixture was then centrifuged (2614 G) for 30 min, after which the solid was precipitated at the bottom of the tube. Aliquots (4 mL) of supernatant liquid were then subjected to rotary evaporation using a Büchi rotary evaporator at 45 °C until at least 95% of extract solvent was removed. The samples were re-suspended in 1 mL of aqueous 3% formic acid solution and then passed through a 0.45 μm nylon filter.

2.3. Standard and Calibration Curves

Commercially available anthocyanin reference standard cyanidin 3-galactoside (2.0 g) was dissolved in 10 mL of 3% formic acid to produce concentrations of 200 μg/mL. A serial dilution was conducted in order to obtain concentrations of 150 μg/mL, 100 μg/mL, 50 μg/mL and 25 μg/mL. Standard solutions were injected separately under direct HPLC conditions in order to generate calibration curves for reference compounds. Standard deviation and percentage RSD was also calculated. The % RSD value for all five concentrations was well below 5%. The average values of all five peaks were plotted against the five concentrations and a linear equation was obtained with an R^2 value of 0.983.

2.4. HPLC Analysis of the Extract Samples

Berry extract samples (15 μL) were analyzed using an Agilent 1100 series HPLC system equipped with a model G1311A Quaternary pump (SL No. DE 40926119). Separation was carried out using a 3.9 mm × 150 mm Symmetry C18 (5 μm) column (SL No. W31761L 008; Agilent technologies, Santa Clara, CA, USA). The system was equilibrated for 20 min at the primary gradient before each injection. Gradients used were the same for all samples with the exception of lingonberry methanol extracts. For most samples the mobile phases used were: (A) 5% formic acid and (B) 100% methanol. Eluent gradient was (a) 0 min—2% B; then gradually (b) 60 min—60% B. For lingonberry methanol extracts the mobile phase was: (A) 0.1% formic acid, and (B) methanol. Eluent gradient was 1% B—0 min-6% 2% B—20 min-12% 3% B—30 min-55%. LC flow rate was one mL/min. Detection of flavonols and anthocyanins was performed with a Diode Array Detector (DAD). We utilized a DAD due to its high specificity and the fact that it has been successfully used to detect the same or similar compounds that we expected to be present in our samples [25,26]. Detection wavelengths for flavonols used were 360 nm and for anthocyanins 520 nm. For our analysis, we utilized similar MS parameters as those developed by Cho et al. [3]. For MS we used the electrospray ionization (ESI) mode. Both positive and negative ionization modes were applied. The parameters were capillary voltage 4.0 kV, nebulizing pressure 30.0 psi, drying gas flow 9.0 mL/min and the temperature was 300 °C. Data collection was performed at 1.0 s per cycle with a mass range of m/z 100—1000. For all extracts, various HPLC chromatograms and their corresponding MS values were compared with values available in the literature for identification purposes.

2.5. Cell Culture

Cortical cell cultures were prepared from neonatal rat pups (1–3 days old) according to previously published methods [27,28]. These procedures were approved by the Institutional Animal Care Committee of Memorial University of Newfoundland. Briefly, dissociated cortices were diluted in serum-containing media [Basal Medium Eagles (GIBCO, Grand Island, NY, USA) containing 10% horse serum (GIBCO), an antibiotic-antimycotic solution at a final concentration of 100 units/mL penicillin G, 100 μg/mL streptomycin sulfate, and 250 ng/mL amphotericin B (Sigma, St. Louis, MO, USA), 0.5% glucose (Sigma), 1 mM sodium pyruvate (GIBCO) and 1% N_2 supplements (GIBCO)] to a concentration of 500,000 cells per mL; cells were then plated in 1-mL aliquots onto collagen-coated six-well FlexPlates (FlexCell, Hillsborough, NC, USA) coated overnight with poly-L-ornithine (500 μg/mL; Sigma). All cultures were maintained in a humidified incubator (5% CO_2, 37 °C). Half of the media in cultures was replaced two days after plating, and then twice per week, with serum-free media containing 2% B27 supplements (GIBCO). Glia formed a confluent monolayer that adhered to the membrane substrate, whereas neurons adhered to the underlying glia. These cultures contained approximately 12% neurons as determined by NeuN, which is a neuronal marker expressed strongly in nuclei and perikarya [29], with the remaining cells representing the glial population. Approximately 95% of NeuN-negative cells stained positively for glial fibrillary acidic protein (Invitrogen, Camarillo, CA, USA), suggesting that the majority of glial cells in these cultures were composed of astrocytes. These cultures were used for experiments at 9–15 days in vitro.

2.6. Cell Injury

Cortical cultures were stretch-injured using a model 94 A Cell Injury Controller developed by Ellis et al. [30] (Bioengineering Facility, Virginia Commonwealth University, Richmond, VA, USA). In brief, the Silastic membrane of the FlexPlate well is rapidly and transiently deformed by a 50-ms pulse of compressed nitrogen or air, which deforms the Silastic membrane and adherent cells to varying degrees controlled by pulse pressure. The extent of cell injury—produced by deforming the Silastic membrane on which the cells are grown—is dependent on the degree of deformation, or stretch. Based on previous work, we used a level of cell stretch equivalent to 5.5 mm deformation (31% stretch), which has been defined as a "mild" level of injury [30,31]. Lingonberry extract (1 μL) was added to the cell cultures (in 1 mL of media) 15 min before injury. One μL of solvent (3% formic acid) was added to some cultures to assess the effect of the solvent alone on cells.

2.7. Cell Counts and Statistical Analysis

Cell cultures were fixed for 20 min with 4% paraformaldehyde as previously described [25,26]. Cultures were then dehydrated with ethanol and mounted with 4′,6-diamidino-2-phenylindole (DAPI), which labels the nuclei of all cells. Images were captured using a Zeiss ObserverA1 microscope and a Pixelfly qe CCD camera (pco., Kelheim, Germany) using AxioVision software (Zeiss). DAPI-positive cells were counted in five separate fields in each culture well at a magnification of 40×. The data for the bioactivity (cell culture) experiments were analyzed with one-way ANOVA ($p < 0.05$) followed by Tukey's multiple comparisons test, and are expressed as % of control values.

3. Results

3.1. Analysis of Lingonberry (V. vitis-idaea) Extracts

Lingonberry samples were treated with extraction solvent [MeOH/H_2O/HCOOH (60:37:3 $v/v/v$)], processed and analyzed by HPLC–MS as described in the materials and methods section. Chemical compounds were identified from the HPLC peaks and are summarized in Table 1. We have used the method described by Cho et al. [3] for extraction of various berry samples as described in the materials and methods section. However, this method did not give satisfactory extraction for all samples. Therefore, for the analysis of some of the lingonberry extracts, the sample was first treated with acidified methanol [MeOH in 1M HCl (85:15 v/v)]. During the analysis the sample: solvent ratio was 1:8. The pH was adjusted to 1.0 (25 °C) using 1M HCl. We used the methodology similar to that developed by Hosseinian and Beta [15]. The sample solvent mixture was sonicated for 30 min and the mixture was transferred to a 50 mL Falcon tube, which was centrifuged at 2614 G for 45 min. Supernatant was collected and subjected to rotary evaporation until 95% of its original volume was removed, which was then spun at 2614 G for another 20 min. The supernatant obtained was filtered using a 0.45 μm nylon filter prior to analysis using the LC–MS instrument. Different compounds using this procedure were identified from the HPLC peaks and are summarized in Table 2. The two different solvent systems were used in order to determine different compounds present in the lingonberry extracts. It was assumed that some compounds which were not extracted in the first solvent system used (MeOH/H_2O/HCOOH) may be isolated in the second solvent system (MeOH/HCl) and more easily detected. This approach isolated additional compounds such as proanthocyanidin A, quercetin-3-glucoside and quercetin-3-O-α arabinoside, which were not detected in the first solvent system.

Table 1. Compounds identified in lingonberry (*V. vitis-idaea*) MeOH/H_2O/HCOOH extracts.

No.	HPLC RT (min)	Identification	*m/z* Values		*m/z* Values in Literature [1]
			[M−H]⁻	[M+H]⁺	[M+H]⁺
1	16.18	Cyanidin-3-glucoside	447.1	449.1	449
2	17.70	Cyanidin-3-galactoside		449.1	449
3	18.54	Cyanidin-3-arabinoside		419.1	419

There were some other unknown peaks in the HPLC chromatogram, which could not be identified. [1] Ek et al., (2006) [22].

Table 2. Compounds identified in lingonberry (*V. vitis-idaea*) methanol extracts.

No.	HPLC RT (min)	Identification	*m/z* Value	*m/z* Value in Literature [1]
		Anthocyanins	[M+H]+	[M+H]+
1	25.62	Cyanidin-3-glucoside	449.2	449
2	25.92	Cyanidin-3-galactoside	449.1	449
3	27.49	Proanthocyanidin A	577.1	577
4	31.34	Quercetin-3-glucoside	465.1	465
5	32.28	Quercetin-3-O-α arabinoside	435.1	435

There were some other unknown peaks in the HPLC chromatogram, which could not be identified. [1] Ek et al., (2006) [22].

3.2. Confirmation and Quantification of Cyanidin-3-Galactoside in Lingonberry Extracts

We chose to quantify cyanidin-3-galactoside in *V. vitis-idaea* as this compound has been reported to be the major anthocyanin in this species [32]. It was found that the retention time (17.70 min) for LC peaks of cyanidin-3-galactoside present in *V. vitis-idaea* extracts was similar to the retention time for LC peaks of the cyanidin-3-galactoside standard, which confirms the presence of this compound in the extracts. From HPLC analysis the area under peak was found to be 3978.62. From the area vs. concentration graph after calculation the concentration of cyanidin-3-galactoside was obtained as 133.71 μg/mL. Therefore, in 100 g of extract the amount of cyanidin-3-galactoside was 66.33 mg or 0.06633 g. Therefore, the concentration of cyanidin-3-galactoside in the *V. vitis-idaea* sample was 0.066%.

3.3. Analysis of Blueberry (V. angustifolium) Extracts

The blueberry extracts were first treated with 40 mL of 2M HCl in ethanol. Hydrolysis was performed at 90 °C in an oil bath for 90 min to break the glycoside linkage. We used a method similar to Nyman and Kumpulainen [20]. Hydrolysis was conducted to produce the flavylium ion (aglycon part), which was identified in the MS system. After hydrolysis, the extract solution was diluted and filtered using a 0.45 μm nylon filter. The sample was then analyzed using a LC–MS instrument. Conditions for MS were the same as those of other analyses. Different compounds identified in blueberry ethanol extracts are summarized in Table 3. The HPLC peaks were identified based on comparison of detected and calculated molecular ions as mentioned in the literature [3,15]. It is to be noted that there is a lack of reproducibility in exact retention times as these may be influenced by aspects such as room temperature, the column and the solvent system. We found very similar retention times when using the same solvent however, so we strongly believe that the different retention times were due to the various solvents.

Table 3. Compounds identified in blueberry (*V. angustifolium*) ethanol extracts.

No.	HPLC RT(min)	Identification	*m/z* Value	*m/z* Value in Literature [1,2]
		Anthocyanins	[M]+	[M]+
1	10.08	Delphinidin-3-galactoside	465.1	465
2	10.52	Delphinidin-3-glucoside	465.1	465
3	10.88	Cyanidin-3-galactoside	449.1	449
4	11.19	Delphinidin-3-arabinoside	435.1	435
5	12.19	Petunidin-3-glucoside	479.1	479
6	13.48	Malvidin-3-glucoside	493.1	493
7	14.14	Peonidin-3-glucoside	463.1	463

There were some other unknown peaks in the HPLC chromatogram, which could not be identified. [1] Cho et al., (2004) [3]; [2] Hosseinian and Beta (2007) [15].

In a separate experiment, blueberry extracts were first treated with acidified methanol (20 mL) [MeOH in 1M HCl (85:15 *v/v*)]. In this experiment the method used was similar to that of Hosseinian and Beta [15]. This experiment was conducted in order to evaluate whether any additional compounds would be detected using acidified methanol rather than acidified ethanol as the solvent due to polarity. Acidified methanol also acts as a very good solvent system for anthocyanin compounds. Various chemical compounds were identified from the LC peaks and are summarized in Table 4.

Table 4. Compounds identified in blueberry (*V. angustifolium*) methanol extracts.

No.	HPLC RT(min)	Identification	*m/z* Value	*m/z* Value in Literature [1,2]
		Anthocyanins	[M]⁺	[M]⁺
1	22.68	Delphinidin-3-galactoside	465.1	465
2	23.71	Delphinidin-3-arabinoside	435.1	435
3	24.70	Cyanidin-3-galactoside	449.1	449
4	25.90	Petunidin-3-galactoside	479.1	479
5	30.24	Malvidin-3-galactoside	493.1	493
6	33.15	Peonidin-3-glucoside	463.1	463
		Flavonols	[M]⁻	[M]⁻
7	20.50	Myricetin-3-rhamnoside	463.0	463
8	22.20	Quercetin-3-galactoside	463.0	463

There were some other unknown peaks in the HPLC chromatogram, which could not be identified. [1] Cho et al., (2004) [3], [2] Hosseinian and Beta (2007) [15].

3.4. Analysis of Black Currant (R. lacustre) Extracts

For the analysis of the black currant extracts, the samples were first treated with acidified methanol. The pH was adjusted to 1.0 (25 °C) and 1M HCl was used for this purpose. Various compounds identified in *R. lacustre* are summarized in Table 5. In a separate experiment the black currant sample was treated with 70% aqueous acetone having 0.01M HCl (20.0 mL) and sonicated for 20 min, as reported by Anttonen and Karjalainen [21]. The compounds identified in black currant extracts by this method are listed in Table 6. The HPLC peaks were identified based on comparison of detected and calculated molecular ions as mentioned in previous reports [3,21].

Table 5. Compounds identified in black currant (*R. lacustre*) methanol extracts.

No.	HPLC RT(min)	Identification	*m/z* Values [M]⁺ or [M+H]⁺	*m/z* Values in Literature[1,2] [M]⁻ or [M−H]⁻
1	1.55	chlorogenic acid	353.0	353
2	10.48	myricetin-3-rhamnoside	463.1	463
3	11.05	quercetin-3-rutinoside	609.2	609
4	12.03	kaempferol rutinoside	595.2	595

There were some other unknown peaks in the HPLC chromatogram, which could not be identified. [1] Anttonen and Karjalainen (2006) [21]; [2] Cho et al., (2004) [3].

Table 6. Compounds detected in black currant (*R. lacustre*) acetone extracts.

No.	HPLC RT(min)	Identification	*m/z* Values [M]⁺ or [M+H]⁺	*m/z* Values in Literature [1,2] [M]⁻ or [M−H]⁻
1	1.45	chlorogenic acid	353.0	353
2	7.45	quercetin glucoside	465.1	465
3	8.57	kaempferol glucoside	449.1	449

There were some other unknown peaks in the HPLC chromatogram which could not be identified. [1] Anttonen and Karjalainen (2006) [21]; [2] Cho et al., (2004) [3].

3.5. Protective Effect of Lingonberry Extract Against Traumatic Injury

In preliminary studies we found that treatment of cultures with 1 μL of solvent alone had no significant change in cell number after 24 h. Injury caused ~30% cell loss by 24 h after injury (Figure 1). Injury in the presence of solvent alone caused a similar loss of cells while the effect of injury was completely reversed in presence of the lingonberry fruit extract (added 15 min before injury). This suggests a high protective effect of lingonberry fruit against traumatic injury in rat brain cultures.

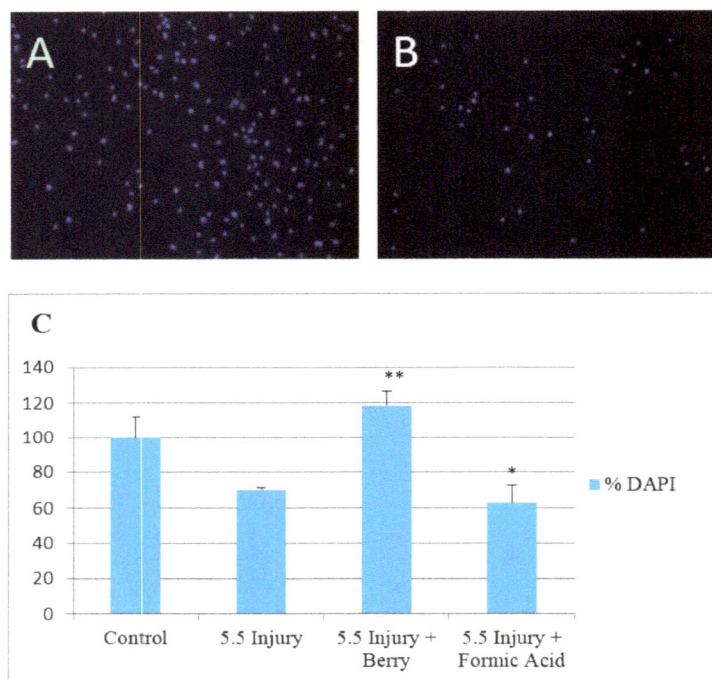

Figure 1. Protective effects of lingonberry extract against in vitro trauma. (**A**) Representative image of DAPI stained nuclei (**blue**) in an uninjured cortical cell well at 9 days in vitro (DIV); (**B**) Representative image of DAPI stained nuclei (**blue**) in an injured cortical cell well (5.5 mm injury) at 9 DIV; (**C**) % of control of DAPI stained nuclei of cortical cells 9–15 DIV under various conditions (*n* = 6 wells in each condition from 3 different culture preparations). Formic acid refers to solvent alone. * statistically different from control group $p < 0.05$. ** Statistically different from 5.5 mm injury group $p < 0.01$. Values are mean ± S.E.M.

4. Discussion

4.1. Comparisons to Findings from Various Berry Species in Other Regions

In our lingonberry extracts, only six compounds were identified (See Table 7). A previous report conducted on *V. vitis-idaea* growing in Finland describes the identification of as many as 28 compounds in methanolic extracts of that species [22]. Comparing HPLC conditions we found that this report used similar analytical conditions as used by our lab. We are unsure of the explanation for this difference in the number of compounds, considering that we obtained excellent peak separation during analysis of the extract and the drying method used to produce the original extract maintains at least 94% (label claimed) of the bioactive compounds intact. It is well known that the chemical composition differs significantly depending on soil, subspecies and climates. As these berries were grown in Newfoundland compared to Finland, it is possible that due to different conditions the number of compounds may vary. We tested levels of cyanidin-3-galactoside in extract samples from *V. vitis-idaea*. We successfully detected a level of 66.33 mg/100 g cyanidin-3-galactoside in Newfoundland *V. vitis-idaea*, which is noteworthy since this compound is considered to have fairly high antioxidant activity [33].

Table 7. Summary of compounds present in the various berry extracts.

Compounds	Blueberry *V. angustifolium*	Black Currant *R. lacustre*	Lingonberry *V. vitis-idaea*
Delphinidin-3-galactoside	X		
Delphinidin-3-glucoside	X		
Delphinidin-3-arabinoside	X		
Cyanidin-3-galactoside	X		X
Cyanidin-3-glucoside			X
Cyanidin-3-arabinoside			X
Petunidin-3-galactoside	X		
Petunidin-3-glucoside	X		
Peonidin-3-glucoside	X		
Malvidin-3-galactoside	X		
Malvidin-3-glucoside	X		
Chlorogenic acid		X	
Myrecitin-3-rhamnoside	X	X	
Quercetin-3-galactoside	X		
Quercetin-3-rutinoside		X	
Kaempferol rutinoside		X	
Quercetin glucoside		X	
Kaempferol glucoside		X	
Proanthocyanidin A			X
Quercetin-3-glucoside			X
Quercetin-3-O-α arabinoside			X

In blueberry powder, a total of 11 compounds were identified (Table 7). Among them, 9 compounds were anthocyanins and two were flavonols. Cho et al. [3] reported that blueberry genotypes including two commercial cultivars found in the United States, Bluecrop (*V. corymbosum* L) known as Northern Highbush and Ozarkblue (a hybrid of majority *V. corymbosum* with some contribution from *V. darrowi* L and *V. ashei*), also known as Southern Highbush, contained 17 and 15 anthocyanins, respectively. This suggests that more compounds may be present in blueberry species found in the United States compared to those of Newfoundland, which may be due to climate, soil and collection time, as the extraction method and analysis conditions were the same. However, our blueberry samples are from a different species (*V. angustifolium*) as compared to the blueberry cultivars analyzed by Cho et al. [3]. Therefore, blueberry cultivars may contain more anthocyanins than naturally growing *V. angustifolium* in Newfoundland. The amount of different anthocyanins detected in our extract powder sample was less than previously reported for *V. angustifolium* and *V. myrtilloides* wild blueberries found in Quebec [16] and Manitoba [15]. This may be due to the fact that those studies were conducted in berry fruits versus extract powders, or to differences in the natural composition of the berries.

In black currant six compounds were identified including chlorogenic acid (3-caffeoylquinic acid), myrecitin-3-rhamnoside, quercetin-3-rutinoside, quercetin glucoside kaempferol rutinoside and kaempferol glucoside (Table 7). This represents a different profile of compounds compared to previous work conducted on black currant (*R. nigrum*) species growing in Finland [20,21] as well as black currant cultivars in Finland [19]. Slimestad and Solheim [34] worked on *R. nigrum* available in Norway and reported that 15 anthocyanins were present in those berries. Different extraction processes and HPLC conditions may be the reason that the number of compounds identified in *R. nigrum* found in Norway was more than in our findings in *R. lacustre*. Then again, the chemical profiles may represent a distinct difference between *R. nigrum* and *R. lacustre*. Määttä et al. [35] identified eight anthocyanins in acidified methanolic extracts of *R. nigrum* available in Sweden. Mikkonen et al. [19] identified four compounds, myricetin, morin, quercetin and kaempferol, in black currant cultivars grown in Finland, which show similarity to our findings.

4.2. Potential Bioactivities of the Analyzed Berry Species

Antioxidant compounds obtained in the diet or through dietary supplementation could be beneficial for combating reactive oxygen and nitrogen species, which can contribute to the development of several disorders, including cancer, cardiovascular disease, diabetes and neurodegeneration [33,36,37]. Our laboratory has a specific interest in identifying natural compounds that may be neuroprotective. We feel that berries or their constituents could potentially be useful for treating or preventing rapid brain aging [38], and may even be useful for decreasing damage associated with more severe ailments such as stroke and traumatic brain injury [39]. However, the extent to which polyphenols can cross the blood brain barrier is largely unknown [40]. Recently, we have found that cortical cells derived from rat brains are protected from glutamate-mediated neurotoxicity by extracts derived from locally collected Newfoundland *V. angustifolium* and *V. vitis-idaea* fruits and leaves [28]. The extracts from blueberry fruits were further tested using the in vitro trauma model described in this paper, and they were found to be highly protective against injury (unpublished data). Interestingly, we found that the leaves of these two species had a higher polyphenolic content and free radical scavenging ability as compared to the fruits. Other studies have also found that the leaves of *V. angustifolium* and *V. vitis-idaea* have a high content of polyphenolic compounds [14,41] suggesting potential sources of dietary supplements.

5. Conclusions

Overall, the results suggest that various species of berries growing natively in Newfoundland contain a wide array of anthocyanins and flavonols with high levels of antioxidant capacity. Further investigation is required to confirm other polyphenols potentially present in these species and to quantify levels of various compounds with antioxidant capacity. In addition, this data demonstrates that the drying technique used to produce these extracts maintains several polyphenolic compounds in the samples, suggesting that these forms of powdered berry extracts could be viable dietary supplements. Further experimentation is aimed at conducting detailed chemical analyses of the leaves of these species, deciphering which species of berries are most neuroprotective, and determining if specific polyphenols are responsible for neuroprotection. In addition, the extent to which polyphenols cross the blood brain barrier is currently being investigated in rodent models.

Acknowledgments: This research was supported by the Natural Sciences and Engineering Research Council of Canada (NSERC), the Canada Foundation for Innovation (CFI) and start-up funding from the School of Pharmacy, Memorial University of Newfoundland.

Author Contributions: J.T.W. and M.D. conceived and designed the experiments; M.Z.H. and E.S. performed the experiments and analyzed the data. J.T.W. and M.D. contributed reagents, materials and analysis tools; M.Z.H. and J.T.W. wrote the paper.

Conflicts of Interest: The authors declare no conflicts of interest.

References

1. Battino, M.; Beekwilder, J.; Denoyes-Rothan, B.; Laimer, M.; McDougall, G.J.; Mezzetti, B. Bioactive compounds in berries relevant to human health. *Nutr. Rev.* **2009**, *67*, S145–S150. [CrossRef] [PubMed]
2. Merken, H.M.; Beecher, G.R. Measurement of food flavonoids by high performance liquid chromatography: A review. *J. Agric. Food Chem.* **2000**, *48*, 577–599. [CrossRef] [PubMed]

3. Cho, M.J.; Howard, L.R.; Prior, R.L.; Clark, J.R. Flavonoid glycosides and antioxidant capacity of various black berry, blueberry and red grape genotypes determined by high-performance liquid chromatography/mass spectrometry. *J. Sci. Food Agric.* **2004**, *84*, 1771–1782. [CrossRef]

4. Moyer, R.A.; Hummer, K.E.; Finn, C.E.; Frei, B.; Wrolstad, R.E. Anthocyanins, phenolics and antioxidant capacity in diverse small fruits: *Vaccinium, Rubus* and *Ribes. J. Agric. Food Chem.* **2002**, *50*, 519–525. [CrossRef] [PubMed]

5. Stintzing, F.C.; Stintzing, A.S.; Carle, R.; Frei, B.; Wrolstad, R.E. Color and antioxidant properties of cyanidin-based anthocyanin pigments. *J. Agric. Food Chem.* **2002**, *50*, 6172–6181. [CrossRef] [PubMed]

6. Prior, R.L.; Wu, X. Anthocyanins: Structural characteristics that result in unique metabolic patterns and biological activities. *Free Radic. Res.* **2006**, *40*, 1014–1028. [CrossRef] [PubMed]

7. Kalt, W.; Dufour, D. Heath functionality of blueberries. *Hortic. Technol.* **1997**, *7*, 216–221.

8. Bravo, L. Polyphenols: Chemistry, dietary sources, metabolism, and nutritional significance. *Nutr. Rev.* **1998**, *56*, 317–333. [CrossRef] [PubMed]

9. Wang, H.; Cao, G.; Prior, R.L. Oxygen radical absorbing capacity of anthocyanins. *J. Agric. Food Chem.* **1997**, *45*, 304–309. [CrossRef]

10. Matsufuji, M.; Otsuki, T.; Takeda, T.; Chino, M.; Takeda, M. Identification of reaction products of acylated anthocyanins from red radish with peroxyl radicals. *J. Agric. Food Chem.* **2003**, *51*, 3157–3161. [CrossRef] [PubMed]

11. Hertog, M.G.; Hollman, P.C.; Venema, D.P. Optimization of a quantitative HPLC determination of potentially anticarcinogenic flavonoids in vegetables and fruits. *J. Agric. Food Chem.* **1992**, *40*, 1591–1598. [CrossRef]

12. Revilla, E.; Ryan, J.-M.; Martin-Ortega, G. Comparison of several procedures used for the extraction of anthocyanins from red grapes. *J. Agric. Food Chem.* **1998**, *46*, 4592–4597. [CrossRef]

13. Chandra, A.; Rana, J.; Li, Y. Separation, identification, quantification, and method validation of anthocyanins in botanical supplement raw materials by HPLC and HPLC-MS. *J. Agric. Food Chem.* **2001**, *49*, 3515–3521. [CrossRef] [PubMed]

14. Harris, C.S.; Burt, A.J.; Saleem, A.; Le, P.M.; Martineau, L.C.; Haddad, P.S.; Bennett, S.A.; Arnason, J.T. A single HPLC-PAD-APCI/MS method for the quantitative comparison of phenolic compounds found in leaf, stem, root and fruit extracts of *Vaccinium angustifolium. Phytochem. Anal.* **2007**, *18*, 161–169. [CrossRef] [PubMed]

15. Hosseinian, F.S.; Beta, T. Saskatoon and wild blueberries have higher anthocyanin contents than other Manitoba berries. *J. Agric. Food Chem.* **2007**, *55*, 10832–10838. [CrossRef] [PubMed]

16. Nicoue, E.E.; Savard, S.; Belkacemi, K. Anthocyanins in wild berries of Quebec: Extraction and identification. *J. Agric. Food Chem.* **2007**, *55*, 5626–5635. [CrossRef] [PubMed]

17. Zheng, Y.; Wang, C.Y.; Wang, S.Y.; Zheng, W. Effect of high-oxygen atmospheres on blueberry phenolics, anthocyanins, and antioxidant capacity. *J. Agric. Food Chem.* **2003**, *51*, 7162–7169. [CrossRef] [PubMed]

18. Zhang, Z.; Kou, X.; Fugal, K.; McLaughlin, J. Comparision of HPLC methods for determination of anthocyanins and anthocyanidins in bilberry extracts. *J. Agric. Food Chem.* **2004**, *52*, 688–691. [CrossRef] [PubMed]

19. Mikkonen, T.P.; Määttä, K.R.; Hukkanen, A.T.; Kokko, H.I.; Torronen, A.R.; Karenlampi, S.O.; Karjalainen, R.O. Flavonol Content Varies among Black Current Cultivars. *J. Agric. Food Chem.* **2001**, *49*, 3274–3277. [CrossRef] [PubMed]

20. Nyman, N.A.; Kumpulainen, J.T. Determination of Anthocyanidins in berries and red wine by high performance liquid Chromatography. *J. Agric. Food Chem.* **2001**, *4*, 4183–4187. [CrossRef]

21. Anttonen, M.J.; Karjalainen, R.O. High performance liquid chromatography analysis of black currant (*Ribes nigrum* L.) fruit phenolics grown either conventionally or organically. *J. Agric. Food Chem.* **2006**, *54*, 7530–7538. [CrossRef] [PubMed]

22. Ek, S.; Kartimo, H.; Mattila, S.; Tolonen, A. Characterization of phenolic compounds from lingonberry. *J. Agric. Food Chem.* **2006**, *54*, 9834–9842. [CrossRef] [PubMed]

23. Kylli, P.; Nohynek, L.; Puupponen-Pimiä, R.; Westerlund-Wikström, B.; Leppänen, T.; Welling, J.; Moilane, E.; Heinonen, M. Lingonberry (*Vaccinium vitis-idaea*) and European cranberry (*Vaccinium microcarpon*) proanthocyanidins: Isolation, identification, and bioactivities. *J. Agric. Food Chem.* **2011**, *59*, 3373–3384. [CrossRef] [PubMed]

24. Nindo, C.I.; Tang, J. Refractance Window Technology: A novel contact drying method. *Dry. Technol.* **2007**, *25*, 37–48. [CrossRef]

25. Omoba, O.S.; Obafaye, R.O.; Salawu, S.O.; Boligon, A.A.; Athayde, M.L. HPLC-DAD Phenolic Characterization and Antioxidant Activities of Ripe and Unripe Sweet Orange Peels. *Antioxidants* **2015**, *4*, 498–512. [CrossRef] [PubMed]

26. Giuffrè, A.M. HPLC-DAD detection of changes in phenol content of red berry skins during grape ripening. *Eur. Food Res. Technol.* **2015**, *237*, 555–564. [CrossRef]

27. Weber, J.T.; Lamont, M.; Chibrikova, L.; Fekkes, D.; Vlug, A.S.; Lorenz, P.; Kreutzmann, P.; Slemmer, J.E. Potential neuroprotective effects of oxyresveratrol against traumatic injury. *Eur. J. Pharmacol.* **2012**, *680*, 55–62. [CrossRef] [PubMed]

28. Vyas, P.; Kalidindi, S.; Chibrikova, L.; Igamberdiev, A.U.; Weber, J.T. Chemical Analysis and Effect of Blueberry and Lingonberry Fruits and Leaves against Glutamate-Mediated Excitotoxicity. *J. Agric. Food Chem.* **2013**, *61*, 7769–7776. [CrossRef] [PubMed]

29. Wolf, H.K.; Buslei, R.; Schmidt-Kastner, R.; Schmidt-Kastner, P.K.; Pietsch, T.; Wiestler, O.D.; Blümcke, I. NeuN: A useful neuronal marker for diagnostic histopathology. *J. Histochem. Cytochem.* **1996**, *44*, 1167–1171. [CrossRef] [PubMed]

30. Ellis, E.F.; McKinney, J.S.; Willoughby, K.A.; Liang, S.; Povlishock, J.T. A new model for rapid stretch-induced injury of cells in culture: Characterization of the model using astrocytes. *J. Neurotrauma* **1995**, *12*, 325–339. [CrossRef] [PubMed]

31. Slemmer, J.E.; Matser, E.J.; De Zeeuw, C.I.; Weber, J.T. Repeated mild injury causes cumulative damage to hippocampal cells. *Brain* **2002**, *125*, 2699–2709. [CrossRef] [PubMed]

32. Lehtonen, H.M.; Rantala, M.; Suomela, J.P.; Viitanen, M.; Kallio, H. Urinary excretion of the main anthocyanin in lingonberry (*Vaccinium. vitis-idaea*), cyanidin 3-O-galactoside, and its metabolites. *J. Agric. Food Chem.* **2009**, *57*, 4447–4451. [CrossRef] [PubMed]

33. De Rosso, V.V.; Morán Vieyra, F.E.; Mercadante, A.Z.; Borsarelli, C.D. Singlet oxygen quenching by anthocyanin's flavylium cations. *Free Radic. Res.* **2008**, *42*, 885–891. [CrossRef] [PubMed]

34. Slimestad, R.; Solheim, H. Anthocyanins from black currants (*Ribes nigrum* L.). *J. Agric. Food Chem.* **2002**, *50*, 3228–3231. [CrossRef] [PubMed]

35. Määttä, K.R.; Kamal-Eldin, A.; Törrönen, A.R. High-performance liquid chromatography (HPLC) analysis of phenolic compounds in berries with diode array and electrospray ionization mass spectrometric (MS) detection: *Ribes* species. *J. Agric. Food Chem.* **2003**, *51*, 6736–6744. [CrossRef] [PubMed]

36. Valko, M.; Leibfritz, D.; Moncol, J.; Cronin, M.T.; Mazur, M.; Telser, J. Free radicals and antioxidants in normal physiological functions and human disease. *Int. J. Biochem. Cell Biol.* **2007**, *39*, 44–84. [CrossRef] [PubMed]

37. Nakamura, T.; Lipton, S.A. Preventing Ca^{2+}-mediated nitrosative stress in neurodegenerative diseases: Possible pharmacological strategies. *Cell Calcium* **2010**, *47*, 190–197. [CrossRef] [PubMed]

38. Lau, F.C.; Shukitt-Hale, B.; Joseph, J.A. The beneficial effects of fruit polyphenols on brain aging. *Neurobiol. Aging* **2005**, *26*, 128–132. [CrossRef] [PubMed]

39. Slemmer. J.E.; Shacka, J.J.; Sweeney, M.I.; Weber, J.T. Antioxidants and free radical scavengers for the treatment of stroke, traumatic brain injury and aging. *Curr. Med. Chem.* **2008**, *15*, 404–414.

40. Slemmer, J.E.; Weber, J.T. Assessing Antioxidant Capacity in Brain Tissue: Methodologies and Limitations in Neuroprotective Strategies. *Antioxidants* **2014**, *3*, 636–648. [CrossRef] [PubMed]

41. Ieri, F.; Martini, S.; Innocenti, M.; Mulinacci, N. Phenolic Distribution in Liquid Preparations of *Vaccinium myrtillus* L. and *Vaccinium vitis idaea* L. *Phytochem. Anal.* **2013**, *24*, 467–475. [CrossRef] [PubMed]

antioxidants

MDPI

Review

Berry Leaves: An Alternative Source of Bioactive Natural Products of Nutritional and Medicinal Value†

Anastasia-Varvara Ferlemi and Fotini N. Lamari *

Laboratory of Pharmacognosy and Chemistry of Natural Products, Department of Pharmacy,
University of Patras, Patras 26504, Greece; abferlemi@upatras.gr
* Correspondence: flam@upatras.gr; Tel.: +30-2610-962335
† Dedicated to the memory of the late Professor Paul Cordopatis, a dedicated scientist and a great mentor.

Academic Editor: Dorothy Klimis-Zacas
Received: 1 February 2016; Accepted: 10 May 2016; Published: 1 June 2016

Abstract: Berry fruits are recognized, worldwide, as "superfoods" due to the high content of bioactive natural products and the health benefits deriving from their consumption. Berry leaves are byproducts of berry cultivation; their traditional therapeutic use against several diseases, such as the common cold, inflammation, diabetes, and ocular dysfunction, has been almost forgotten nowadays. Nevertheless, the scientific interest regarding the leaf composition and beneficial properties grows, documenting that berry leaves may be considered an alternative source of bioactives. The main bioactive compounds in berry leaves are similar as in berry fruits, *i.e.*, phenolic acids and esters, flavonols, anthocyanins, and procyanidins. The leaves are one of the richest sources of chlorogenic acid. In various studies, these secondary metabolites have demonstrated antioxidant, anti-inflammatory, cardioprotective, and neuroprotective properties. This review focuses on the phytochemical composition of the leaves of the commonest berry species, *i.e.*, blackcurrant, blackberry, raspberry, bilberry, blueberry, cranberry, and lingonberry leaves, and presents their traditional medicinal uses and their biological activities *in vitro* and *in vivo*.

Keywords: Vaccinium; Ribes; Rubus; traditional use; polyphenols; chlorogenic acid; analysis

1. Introduction

The everlasting quest for health promoting and disease preventing agents in the developed world has changed our view of food sources; superfoods, functional foods, food supplements, and nutraceuticals were introduced and have enriched the products of the food industry contributing to its further growth [1]. Berry fruits constitute a large group of functional food or "superfoods", whose consumption delivers several health benefits beyond basic nutrition, but little is known about the leaves of berry plants. In this review, we present the phytochemical composition of the leaves of common berry species, as well as summarize their traditional medicinal uses and the results of the evaluation of their biologic properties *in vitro* and *in vivo* so far.

In order to compile the review, a search was performed in the PubMed (http://www.ncbi.nlm.nih.gov/pubmed) and Google Scholar databases. The search terms included the keywords: berries, Vaccinium; Ribes; Rubus, and the common names of each species (the headings of all chapters of this review) and leaves. In January 2016, this search yielded about 500 results in the PubMed database. Information on the traditional uses has also been acquired from the European Medicines Agency (EMA) monographs on the respective herbal medicines (http://www.ema.europa.eu). Original research and review articles were taken into account and special emphasis was placed on published literature concerning the few available clinical studies.

The small edible, brightly colored berries are low energy density fruits, rich in vitamins, fibers, and various phenolic compounds [2]. The edible berries belong to the genera of *Vaccinium* (blueberries, cranberries, bilberries, lingonberries), *Ribes* (gooseberries, black and red currants), *Rubus* (raspberries, blackberries and cloudberries), *Fragaria* (strawberries), *Aronia* (chokeberries), and *Sambucus* (elderberries). All of them contain high levels of phenolics, which greatly contribute to their organoleptic properties and health benefits.

Berry phenolics represent a diverse group of compounds including phenolic acids (hydroxybenzoic and hydroxycinnamic acids, and their derivatives), flavonoids, such as flavonols, flavanols, and anthocyanins, and tannins (gallotannins and ellagitannins), divided into condensed tannins (proanthocyanidins) and hydrolysable tannins. A greater variety of compounds is recorded for blackberries and raspberries of the genus *Rubus*, whereas all of the other berry species are usually characterized by the high levels of specific phenolic groups, *i.e.*, anthocyanins and proanthocyanidins (Figure 1). The berries of genus *Rubus*, as well as the chokeberries, are often richer than other berries in *p*-hydroxybenzoic acids; they also contain moderate levels of hydroxycinnamic acids; however, trace amounts were detected in cloudberries [3–6]. In contrast, blueberries (*Vaccinium* spp.) are the richest source of hydroxycinnamic acids, such as *p*-coumaric acid, chlorogenic acid, and other caffeic acid derivatives [3,4,6–8]. With respect to flavonoids, chokeberries, highbush blueberries, American cranberries (*V. macrocarpon*), blackcurrants, and lingonberries contain the highest concentration of flavonols, especially quercetin and myricetin derivatives and aglycones [3,5–8]. On the contrary, raspberries, cloudberries, red currants, and gooseberries contain only traces of flavonols [4,5].

The bright color of the small edible berries is attributed to the significant quantities of anthocyanins, which are distributed mainly to the epidermal tissues in fruits. Substantial amounts of different types of anthocyanins (glycosylated or not) are found in chokeberries, bilberries, wild and cultivated blueberries, elderberries, blackcurrants, and the European cranberries (*V. oxycoccus*). Bilberries, for example, contain fifteen different anthocyanins, *i.e.*, delphinidin and cyanidin monoglycosides (the principal anthocyanin), petunidin, peonidin, and malvidin glycosides. Strawberries contain mainly pelargonidins, while lingonberry, red currant, chokeberry and elderberry anthocyanins consist only of cyanidin glycosides. Cyanidins are also the principal anthocyanins in American cranberries, while in European cranberries the most abundant are peonidins [4–6,8–10].

Proanthocyanidins, consisting only of procyanidins, *i.e.*, (+)-catechin and (−)-epicatechin polymers, are present in high concentration in chokeberries, high- and lowbush blueberries, American cranberries, and lingonberries [4,11]. Finally, ellagic acids and ellagitannins are in high amounts in the berries belonging to the genus *Rubus* (cloudberries, raspberries), as well as in strawberries [5,10]. Figure 1 summarizes the phenolic content of the most common berry fruits.

A strong body of scientific research documents the contribution of the consumption of berries to the three targets of functional foods: (i) health maintenance (e.g., mental health, immune function); (ii) reduced risk of obesity; and (iii) reduced risk of chronic diet-related diseases (e.g., cardiovascular disease, type 2 diabetes, and metabolic syndrome) [1]. However, not only the fruits, but also the leaves, of the berry plants have been used in traditional remedies; leaf extracts have often been used against several diseases, such as colds, inflammation of the urinary tract, diabetes, and ocular dysfunction by Native Americans and other populations, but these treatments have been almost forgotten nowadays. In the last five years, the European Medicines Agency (EMA) has approved the circulation of leaf infusions and extracts of *Ribes nigrum*, *Rubus idaea*, and *Arctostaphylos uva-ursi* as herbal medicinal products based on their traditional uses and another monograph for the wild strawberry (*Fragaria vesca* L.) leaf extracts has just been announced [12–15].

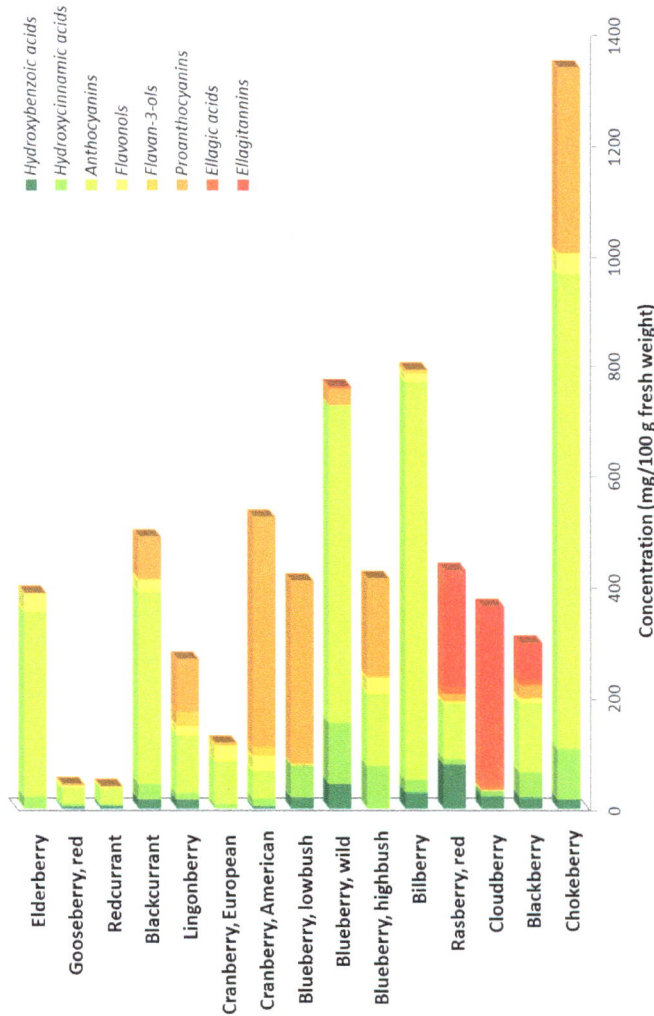

Figure 1. The phenolic composition of the commonest berries. Chokeberries (*Aronia mitschurinii*) contain the highest concentration of bioactive compounds, mainly anthocyanins and proanthocyanins. Bilberry (*Vaccinium myrtillus*) is rich in anthocyanins, as well as wild blueberry (*Vaccinium corymbosum*), which also contains notable amount of phenolic acids. Lowbush blueberry (*Vaccinium angustifolium*) and the American cranberry (*Vaccinium macrocarpon*) are sources of proanthocyanidins. The berries of the genus *Rubus*, *i.e.*, raspberry (*Rubus idaeus*), cloudberry (*Rubus chamaemorus*), and blackberry (*Rubus fruticosus*) contain all the principle bioactive compounds that we meet in berries, and especially ellagitannins. Gooseberry (*Ribes uva-crispa*) and red currant (*Ribes rubrum*), on the other hand, contain mainly phenolic acids and only traces of the other compounds.

Despite their medicinal value, which stems in large from their phenolic/polyphenolic content, berry leaves are the main byproducts of harvesting, meaning that tons of leaves are wasted annually. Analytical studies show that the leaf phenolic composition is similar to that of the precious fruits or even richer and higher, indicating that they may be utilized as an alternative source of bioactive natural products for the development of food supplements, nutraceuticals, or functional foods. This review presents our knowledge heretofore; we present the phytochemical composition of the leaves of the common berry species, as well as summarize the studies of the beneficial activities of their extracts pertaining to their nutritional or medicinal value. The compositions of the berry leaves are summarized in Table 1. The structures of the commonest phenolic acids and derivatives are presented in Figure 2 and of the flavonoid aglycons and terpenes in Figure 3. The traditional medicinal uses and the relevant biological properties demonstrated by *in vitro*, *in vivo*, and clinical studies are presented in each subsection, but also in tabulated form (Table 2).

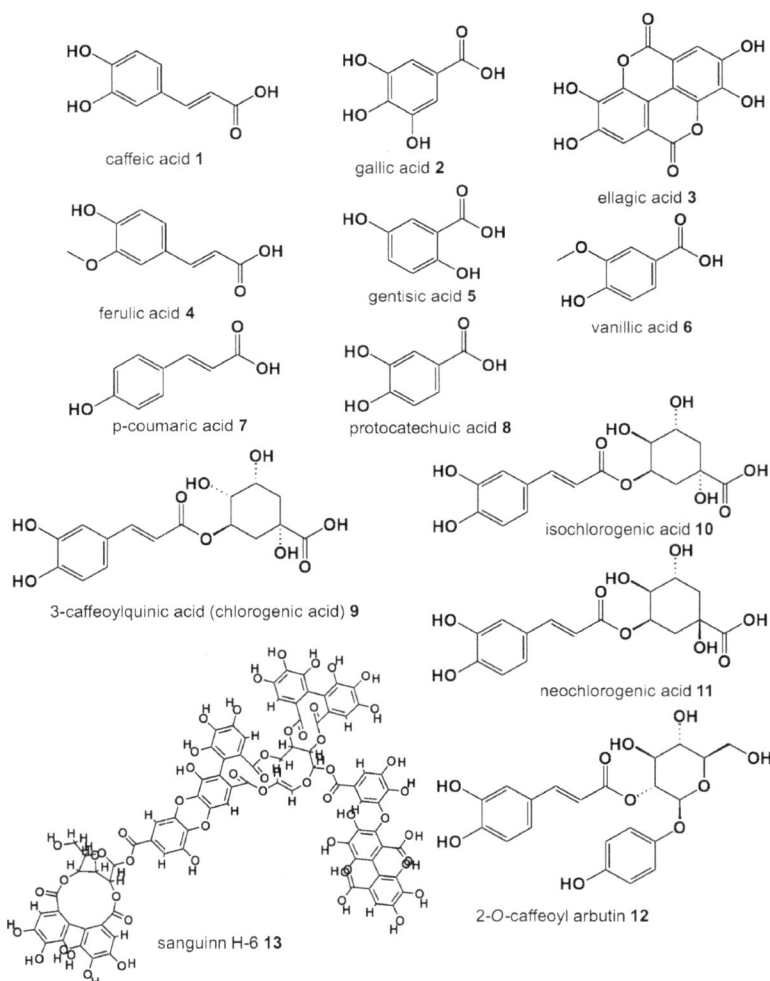

Figure 2. Structures of the most common phenolic acids and acid derivatives of berry leaves.

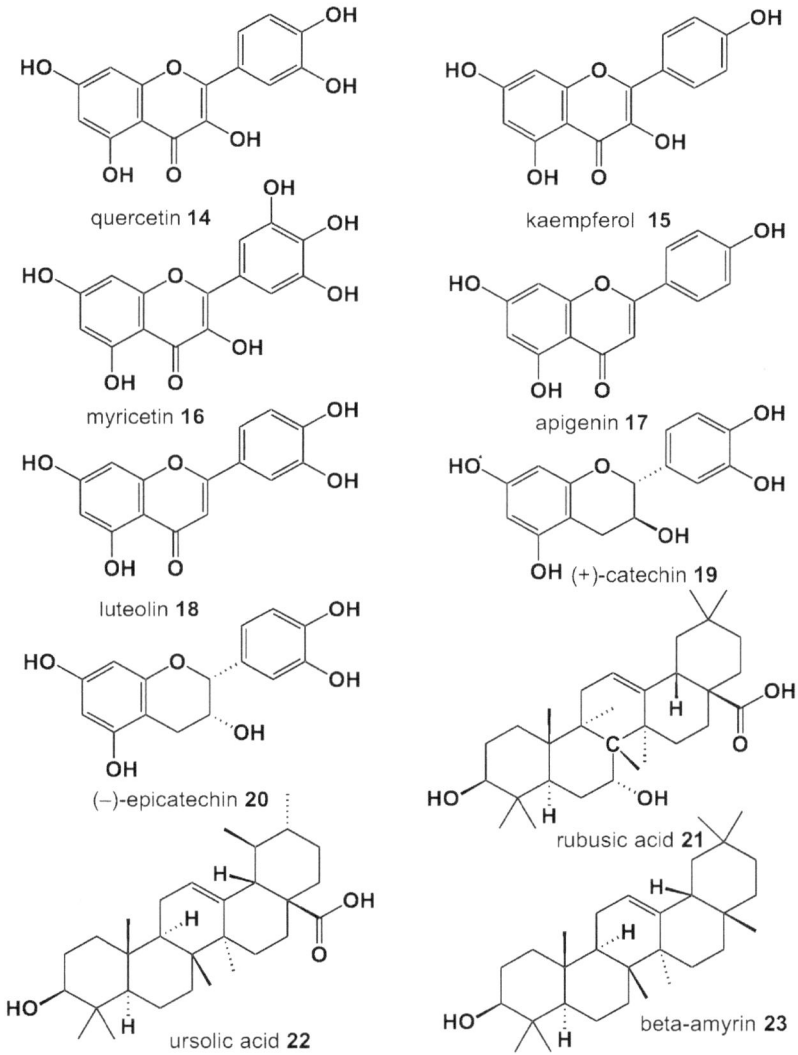

Figure 3. Structures of the main flavonoid aglycons and terpenes of berry leaves.

Table 1. An overview of the distribution of phenolic acids, flavonols, flavan-3-ols, ellagitannins (ET), proanthocyanidins (PACs), anthocyanins in blackcurrant (BC), blackberry (BB), raspberry (RP), bilberry (BI), highbush blueberry (H-BL), lowbush blueberry (L-BL), cranberry (CB) and lingonberry (LB) leaves.

	Compound Name	Berry Leaves							
		BC	BB	RB	BI	H-BL	L-BL	CB	LGB
Phenolic Acids	Chlorogenic acid	BC	BB	RB	BI	H-BL	L-BL	CB	LGB
	Neo-chlorogenic acid	BC	BB			H-BL	L-BL	CB	
	Iso-chlorogenic acid	BC				H-BL			
	Caffeic acid	BC	BB	RB	BI	H-BL	L-BL		LGB
	Gallic acid	BC	BB						
	Ferulic acid	BC				H-BL			
	Quinic acid	BC				H-BL			
	p-coumaric acid	BC	BB	RB	BI	H-BL			LGB
	Coumaroyl-quinic acid						L-BL	CB	LGB
	Caffeoyl-shikimic acid/ Ferroyl-quinic acid isomer				BI				LGB
	Gentisic acid	BC							
	p-hydroxybenzoic acid/ vanillinic acid			RB					LGB
	2-O-caffeoylarbutin								LGB
	Coumaroyl/caffeoyl-hexose hydroxyphenols								LGB
	Ellagic acid		BB	RB	BI	H-BL	L-BL		
Flavonols	Quercetin	BC	BB	RB	BI	H-BL	L-BL		LGB
	Quercetin-3-O-rutinoside	BC	BB	RB		H-BL			LGB
	Quercetin-3-O-galactoside	BC	BB	RB	BI	H-BL	L-BL	CB	LGB
	Quercetin-3-O-glucoside	BC	BB	RB	BI	H-BL	L-BL		LGB
	Quercetin-3-O-glucuronide		BB	RB	BI				
	Quercetin-3-O-a-L-rhamnoside					H-BL	L-BL	CB	LGB
	Quercetin-3-O-(4''-HMG)-a-rhamnoside				BI				LGB
	Quercetin-3-O-arabinoside				BI	H-BL	L-BL		LGB
	Quercetin-3-O-xyloside					H-BL			LGB
	Quercetin-3-O-malonylglucoside	BC						CB	
	Quercetin-3-O-glucosyl-6''-acetate					H-BL		CB	
	Quercetin-3-O-R-arabinofuranoside		BB						LGB
	Kaempferol		BB			H-BL			
	Kaempferol-3-O-rutinoside	BC				H-BL	L-BL		
	Kaempferol-3-O-galactoside	BC							
	Kaempferol-3-O-glucoside	BC		RB					
	Kaempferol-3-O-glucuronide				BI	H-BL			

Table 1. *Cont.*

Compound Name	Berry Leaves							
	BC	BB	RB	BI	H-BL	L-BL	CB	LGB
Flavonols								
Kaempferol-3-O-glucosyl-6''-acetate	BC							
Kaempferol-3-O-malonylglucoside	BC							
kaempferol-(HMG)-rhamnoside								LGB
Kaempferol-pentoside								LGB
Myricetin				BI	H-BL			
Myricetin-3-O-malonylglucoside	BC				H-BL			
Myricetin-3-O-rutinoside/-3-O-galactoside/-3-O-glucoside/-3-O-arabinoside/-3-O-xyloside								
Isorhamnetin-3-O-rutinoside/-3-O-glucoside	BC							LGB
Flavan-3-ols								
Catechin	BC	BB		BI		L-BL		LGB
Epicatechin	BC	BB		BI		L-BL		
Epigallocatechin/ Gallocatechin and its isomers	BC			BI				
Epicatechin gallate methyl gallate		BB	RB					
ETs								
Sanguiin H-6 /Lambertianin C		BB	RB					
Lambertianin D			RB					
Casuarinin		BB						
PACs								
Cinchonains				BI		L-BL		LGB
Proanthocyanidin A1							CB	LGB
Proanthocyanidin A2								
Proanthocyanidin B				BI				
Kandelin A1/A2						L-BL		
Procyanidins/Prodelphinidins						L-BL		
Anthocyanins								
Delphinidin-3-O-glucoside/-3-O-rutinoside	BC							
Cyanidin-3-O-glucoside	BC	BB			H-BL			
Cyanidin-3-O-rutinoside	BC							
Cyanidin-3-O-arabinoside/-3-O-glucuronide					H-BL			

BC: Blackcurrant, BB: Blackberry, RB: Rasberry, BI: Bilberry, H-BL: Highbush blueberry, L-BL: Lowbush blueberry, CB: Cranberry, LGB: Lingonberry, ETs: Ellagitannins, PACs: Proanthocyanidins

Table 2. Medicinal uses and biological properties of berry leaves.

	Blackcurrant	Red Raspberry	Blackberry
EMA	*Traditional Medicinal Product* Minor articular pain Adjuvant in minor urinary complaints [13]	*Traditional Medicinal Product* Symptomatic relief of minor spasm associated with menstrual periods Symptomatic treatment of mild inflammation in the mouth or throat Symptomatic treatment of mild diarrhea [16]	
Traditional uses	Diaphoretic and diuretic agent Against diarrhea Against spasmodic cough Relief of rheumatic pain [14,15]	Labor stimulator [17] Relief of menstrual cramps Relief of diarrhea Astringent agent Anti-inflammatory agent (mouth, throat) Against chronic skin conditions Treatment of conjunctivitis [18]	Mouthwash against thrush, gum inflammations, mouth ulcers, sore throat Against respiratory problems Astringent agent Regulation of anemia, diarrhea, dysentery, cystitis, hemorrhoids [19]
In vitro/In vivo	Antioxidant, Anti-inflammatory activity [14,20] Analgesic activity [15]	Antioxidant activity [21]	Antidiabetic/Hypoglycemic activity [22–24] Antimicrobial activity [25] Analgesic, Anti-inflammatory, Angiogenic activity [19,26,27]
Clinical trials		Indications that it facilitates labor [28–30]	

	Bilberry	Blueberry	Cranberry	Lingonberry
Traditional uses	Diuretic, astringent and antiseptic agent for the urinary tract Antibacterial Anti-inflammatory Antidiabetic [31]	Antidiabetic agent [32,33]		Diuretic agent Antiseptic in urinary tract [31]
In vitro/In vivo	Antidiabetic activity [34–36] Anti-hyperlipidemic activity [37] Antistaphylococcal activity [38] Antioxidant, Anti-neoplastic activity [39]	Antioxidant, Anticataract [40] Neuroprotective activity [41] Antidiabetic activity [42] Antimicrobial activity [43]	Antioxidant activity [44]	Antitussive, Anti-inflammatory Anti-catarrhal activity [45] Neuroprotective activity [46]
Clinical trials			Antimicrobial agent—urinary tract protection Antioxidant activity [47]	

2. Blackcurrant (*Ribes nigrum*) Leaves

In 1996, French Pharmacopoeia included, for the first time, a monograph on *Ribes nigrum* L. folia [48] and in 2010, EMA issued the official community monograph for blackcurrant leaves [14].

The slightly wrinkled leaf is dark green at the upper surface and pale greyish green at the lower surface, on which a widely spaced reticulate venation is particularly distinct. Moreover, the leaves have glands that can be seen as scattered yellowish dots. In contrast with the fresh leaves that are strongly aromatic, the dried leaves have no odor or taste. The leaves should be collected during or shortly after flowering [48].

The phytochemical analysis based on spectroscopic and separation techniques revealed that the most abundant secondary metabolites of blackberry leaf can be subdivided into three groups: hydroxycinnamic acids, flavonoids, and proanthocyanidins. In detail, experiments with thin layer chromatography (TLC) by Trajkovski in 1974 revealed that chlorogenic acid (9) and its isomers (iso- and neo-chlorogenic acid, 10 and 11) are found in relatively high amounts in the foliage of blackcurrant [49]. Other hydroxycinnamic acids in these leaves are caffeic acid (1), gallic acid (2), ferulic acid (4), coumaric acid (7), and gentisic acid (5) [49]. In a recent HPLC-DAD analysis of the ethanolic extract of blackcurrant leaves, only chlorogenic acid and its isomer neo-chlorogenic acid were quantified; their concentrations ranged from 0.081 to 0.121 mg/g dry weight and from 0.044 to 0.435 mg/g dry weight, respectively [50]. In an ethanolic extract of blackcurrant leaves from Estonia, chlorogenic acid concentration was found much higher (14.93 mg/g dried leaves) by HPLC-MS analysis [20]. With regards to flavonols, kaempferol (15) and quercetin (14) derivatives have been reported in the foliage of *Ribes nigrum* [49]. Particularly, Vagiri *et al.* [50] have identified quercetin-3-*O*-rutinoside (0.099–0.229 mg/g dry weight), quercetin-3-*O*-galactoside (0.057–0.081 mg/g dry weight), quercetin-3-*O*-glucoside (0.038–0.085 mg/g dry weight), and the most abundant flavonoid, quercetin-3-*O*-malonylglucoside (2.424–3.890 mg/g dry weight). Additionally, kaempferol-3-*O*-rutinoside (0.019–0.036 mg/g dry weight), kaempferol-3-*O*-glucoside (0.017–0.031 mg/g dry weight), kaempferol-3-*O*-malonylglucoside (0.135–0.409 mg/g dry weight), an isomer of the latter (0.488–2.441 mg/g dry weight), as well as myricetin-malonylglucoside (0.042–0.055 mg/g dry weight) and its isomer (0.019–0.023 mg/g dry weight) were quantified [50]. The range of values depends on the harvest time [50]. The HPLC-MS analysis of the Estonian blackcurrant leaf extract showed that quercetin glucoside (quercetrin) and rutinoside (rutin), the two major glycosides of the ethanolic extract, were found in high concentrations (19.47 and 3.99 mg/g dried leaves, respectively) [20]. Furthermore, isorhamnetin-3-*O*-rutinoside, isorhamnetin-3-*O*-glucoside, kaempferol-3-*O*-galactoside and kaempferol-3-*O*-glucosyl-6"-acetate have been recorded in the ethanolic extract of *Ribes nigrum* leaves [19,51]. Catechin (19), epigallocatechin, and epicatechin (20) have also been detected in the leaves even though not constantly; the concentration of catechin in the ethanolic extract of Estonian leaves was 7.89 mg/g [20]. In a methanolic extract of blackcurrant leaves, Tits *et al.* [52,53] have identified, via high performance thin layer chromatography (HPTLC), nuclear magnetic resolution (NMR), and infrared spectroscopy (IR), several different tannins (catechin, epicatechin, gallocatechin, epigallocatechin, gallocatechin-(4a-8)-gallocatechin, gallocatechin-(4a-8)-epigallocatechin, and the new trimer: gallocatechin-(4a-8)-gallocatechin-(4a-8)-gallocatechin, gallocatechin-(4a-8)-catechin, gallocatechin-(4-6)-gallocatechin, and catechin-(4a-8)–gallocatechin-(4a-8)-gallocatechin).

A significant difference between fruits and leaves is the lowest concentration of anthocyanins in the latter; delphinidin-3-*O*-glucoside and –rutinoside, as well as cyanidin-3-*O*-glucoside and –rutinoside have been identified, but not quantified, in the ethanolic extract of blackcurrant leaves [51]. Another difference is the absence of hydroxybenzoic acids. However, the blackcurrant leaves contain significant substances, such as glycerolipids (mainly alpha-linoleic acid) and ascorbic acid (1–2.70 mg/g dried material) [54]. Finally, the essential oil of the leaves of *Ribes nigrum* contains mainly monoterpenic substances (α-pinene, myrcene, *p*-cymene, limonene, β-ocimene, *etc.*), the sesquiterpenes caryophylene and humulene, as well as methyl salicylate [55].

According to the EMA monograph on *Ribes nigrum* folium, its tea is a traditional medicinal product for minor articular pain and a diuretic to achieve flushing of the urinary tract as an adjuvant in minor urinary complaints [14]. The leaves of blackcurrant have been used in folk medicine for their diaphoretic properties, as well as against diarrhea and spasmodic cough [20,54]. The leaves have significant antioxidant and anti-inflammatory properties (inhibition of myeloperoxidase activity and reactive oxygen species production on activated neutrophils), as it has been demonstrated by biologically-relevant cellular models, which substantiate their traditional uses against inflammatory conditions [54,56]. These effects are correlated with the phenolic ingredients that are originally synthesized by plants to protect themselves from pathogens [57]; therefore, it has been proposed that *Ribes nigrum* leaves might be used for large scale extractions of antioxidant molecules [56]. Furthermore, evidence from carrageenan-induced rat paw edema studies has revealed the potential analgesic properties of blackcurrant leaves, which was later reinforced by findings of blackcurrant proanthocyanidin inhibition of leukocyte infiltration [20].

3. Blackberry (*Rubus fruticosus*) Leaves

Blackberries are perennial shrubs, lasting for three seasons or more. The upper side of the leaves is dark green, while the underside is lighter green. Short prickles cover the stalks and veins of the leaves [25].

Many phytochemical investigations have proven the presence of diverse secondary metabolites in blackberry leaves. In general, they are rich in tannins and they, also, contain a notable amount of flavonoids, phenolic acids, triterpenes, mineral salts, and vitamin C [22,23,25]. More specifically, phenolic acids like ellagic (3), gallic (2), caffeic (1), and *p*-coumaric (7) acids, flavonoids, such as quercetin (14), hyperoside, kaempferol (15), myricitin, (+)-catechin (19), (−)-epicatechin (20), epicatechin gallate, and proanthocyanidin B1 have been identified in the leaves of *R. fruticosus*, as well as in the fruit [25]. The HPLC analysis of a hydrolyzed methanolic extract of blackberry leaves showed that total flavonoids, expressed as quercetin equivalents, range from 0.14% to 0.31% of dry weight, while total ellagic acid ranges from 2.93% to 4.32% of dry weight [24].

In 2015, Ozmianki *et al.* [22] extensively analyzed the phenolic composition of twenty-six different wild blackberry leaf samples by LC/MS QTOF; 33 compounds were detected in the respective methanolic extracts, *i.e.*, 15 flavonols, 13 hydroxycinnamic acids, three ellagic acid derivatives, and two flavones. The total content of phenolic compounds extracted from the leaves of wild blackberries, calculated as the sum of compounds resulting from UPLC-PDA analysis, was highly diverse and ranged from 83.02 mg/g dry matter for *R. austroslovacus* to 334.24 mg/g dry matter for *R. perrobustus*. The largest group of phenolic compounds was that of ellagitannins (51.59–251.01 mg/g dry matter), as in blackberry fruits [52]; the most abundant ellagitannins in the wild blackberry leaves were sanguiin H-6 (13) (range 0–73.92 mg/g dry matter), lambertianin C (range 16.75–123.41 mg/g dry matter), and casuarinin (34.47–117.86 mg/g dry matter). In the same study, the second group of bioactive compounds in the leaves of wild blackberries was that of derivatives of quercetin (14), kaempferol (15), luteolin (18), and apigenin (17) (average content 35.17 mg/g dry matter); kaempferol-3-*O*-glucuronide and quercetin-3-*O*-glucuronide (9.23 and 7 mg/g dry matter, respectively) were the most abundant compounds. The next group of compounds in blackberry leaves is composed of phenolic acids, especially derivatives of caffeic acid (1), *p*-coumaric acid (7), and ellagic acid (3) (average content 28.74 mg/g dry matter); *p*-coumaric acid derivatives and neo-chlorogenic acid (11) were found in notable amounts in blackberry leaf extracts [22].

Robinson *et al.* [26], in 1931, reported the presence of cyanidin-3-*O*-saccharide in blackberry leaves. Almost 40 years ago, two triperpene acids were isolated from *R. fruticosus* leaves, rubusic (21) and rubinic acids; 2-a-hydroxyursolic acid and β-amyrin (23) were also detected [18,27].

Health-promoting effects and immunity-boosting properties have been attributed to blackberry leaves since long ago. Hippocrates recommended blackberry stems and leaves soaked in white wine for facilitating childbirth [16]. Zia-UI-Haq *et al.* [25], in their review of the traditional uses of *Rubus fruticosus* leaves, reported that the decoction of the leaves has been used as tonic and a mouthwash; gargles help treating thrush, gum inflammation, sore throat, and mouth ulcers. The leaves are also chewed in order to strengthen the gums and to cure thrush. A poultice of the leaves is applied to abscesses and skin ulcers as an astringent. In addition, blackberry leaves and roots are a long-standing home remedy for anemia and menses, diarrhea, dysentery, cystitis, and hemorrhoids. Finally, they have traditionally been used against several respiratory problems [19].

Indeed, it has been demonstrated that the leaves of blackberry possess significant antimicrobial activity, higher than the fruit, against several bacterial strains, such as *Salmonella typhi*, *Escherichia coli*, *Staphylococcus aureus*, *Micrococcus luteus*, *Proteus mirabilis*, *Bacillus subtilis*, *Citrobacteri* sp., and *Pseudomonas aeruginosa* [25]. In contrast, when the methanolic extracts of blackberry leaves were tested for their antifungal potential against nine pathogenic fungal strains (*Yersinia aldovae*, *Aspergillus parasiticus*, *Candida albicans*, *Aspergillus niger*, *Aspergillus effusus*, *Macrophomina phaseolina*, *Fusarium solani*, *Trichophyton rubrum*, *Saccharomyses cerevisiae*) they did not have any biological activity [17].

Several studies point out the anti-diabetic effect of blackberry leaf extracts; water and butanol extracts were reported to be active in non-insulin dependent diabetes and had significant hypoglycemic effect in normal rats [58]. Similar results were obtained for the infusion of blackberry leaves in alloxan-diabetic rabbits [59]. Moreover, a tea made from *R. fruticosus* leaves decreased diabetic symptoms (hyperglycemia), a property partly attributed to their content in chromium and zinc [60]. Finally, the antioxidant and angiogenic activities of different extracts of blackberry leaves have also been recorded in several studies [22,28,29].

4. Raspberry (*Rubus idaeus*) Leaves

The green leaves of *Rubus idaeus* have been included in British Pharmacopoeia since 1983 [30] and in 2012 the European Medicines Agency issued a community herbal monograph on red raspberry leaves [13].

The beneficial medicinal properties are attributed to the bioactive compounds of the leaves, which are mainly hydrolysable tannins [21]. Gudej [24] reported that tannin concentration in the dried raspberry leaf ranges from 2.6% to 6.9% (w/w) and that the principle compounds are ellagic acids. Additional ellagitannins that have been identified in these leaves are the dimers sanguiin H-6 (13) and H-10, and the trimers lambertianin D and lambertianin C, as well as methyl gallate [19,21]. The second most abundant group in raspberry leaves is flavonoids. The quantity of flavonoids in the leaves of *R. idaeus* is significantly higher than that in the fruits where flavonoids compose only a very small fraction of the bioactive compounds; leaf flavonoids range from 0.46% to 1.05% (w/w) [31]. In the study of Ozmianski *et al.* [19] the flavonoid fraction was the main phenolic group, constituting almost 11% of leaf extract powder weight.

Phenolic acids, other than ellagic acid (3), have been found in very small amounts, mainly caffeic (0.55 mg/g dried leaf) (1) and chlorogenic acid (0.70 mg/g dried leaf) (9) [61]. Moreover, *p*-coumaric (7), ferulic (4), protocatechuic (8), gentisic (5), caffeoyltartaric, feruloyltartaric, and *p*-coumaroyl-glucoside acids, as well as *p*-hydroxybenzoic and vanillic acids have been reported in raspberry leaves [62]. Finally, terpenoids have been identified, including mono- and sesquiterpenes, like terpinolene and 3-oxo-α-ionol, as well as triterpenes, such as a- and β-amyrin (23), squalene and cycloartenol [6,21].

The study of Gudej [24] presents an interesting comparison of the main *Rubus* categories, *i.e.*, blackberries and raspberries. The leaves of wild raspberry (*R. saxatilis*), cultivated raspberry species (*R. idaeus* Malling Promise) and blackberry (*R. fruticosus* Gazda) had the highest flavonoid content as measured by HPLC. Furthermore, the leaves of raspberries are characterized by lower amounts of both tannins and ellagic acid (3.25% and 2.53% of dry weight respectively) than blackberry leaves (6.50% and 4.32% of dry weight respectively) [24].

Raspberry leaf has been used in Europe for various gynecological disorders, *i.e.*, menstruation, labor and ailments of the gastrointestinal tract (diarrhea) [21,30]. It is reported that a hot tea made from raspberry leaves stimulates and facilitates labor and shortens its duration [21,63]. Other traditional uses include its use as an astringent gargle and less often for chronic skin conditions and for the treatment of conjunctivitis [30]. The European monograph on raspberry leaf has approved its use as a traditional herbal medicinal product for the symptomatic relief of minor spasms associated with menstrual periods, for the symptomatic treatment of mild inflammation in the mouth or throat, and of mild diarrhea [13].

Since raspberry leaf is a commonly used herb during pregnancy today, earlier and current investigations have explored its effects mainly regarding labor. Jing *et al.* [64] reported that pretreatment of pregnant rats with tea did not alter the ability of oxytocin to initiate contractions. Additionally, in pregnant animals treated with red raspberry leaf tea, labor was not augmented by a direct effect on uterine contractility; in contrast, it had variable effects on preexisting oxytocin-induced contractions, sometimes augmenting the effect of oxytocin and sometimes causing augmentation followed by inhibition [64]. Furthermore, these effects depended on the herbal preparation used and on pregnancy status [64].

Two different clinical studies were performed in order to assess the efficacy of raspberry leaf preparations in pregnancy [63,65]. About 150 women were included in the studies. No clinically significant differences were observed among the different groups regarding maternal blood loss, maternal diastolic blood pressure pre labor or transfer to special care baby unit, length of gestation, the likelihood of medical facilitating of labor, and need for pain relief during labor. In addition to these studies, others report that raspberry leaves possess significant antioxidant activity, stronger than the respective extracts of blackberry leaves [29].

5. Bilberry (*Vaccinium myrtillus*) Leaves

Bilberries are one of the most important wild berries in Northern Europe, commonly called European blueberries to distinguish them from the other blueberries. Qualitative and quantitative analysis studies based on LC/MS conclude that the main bioactive compounds of bilberry leaves are hydroxycinnamic acids (36% of the weight of leaf extract powder) and especially chlorogenic acid (9) [19,34,66]; its concentration ranges from 59% to 74% of the total hydroxycinnamic acids [66]. Jaakola *et al.* [37] and Riihinen *et al.* [35] reported that the concentration of hydroxycinnamic acids were higher in the leaves than in the bilberry fruit. Hokkanen *et al.* [34] analyzed a methanolic extract of bilberry leaves by LC/TOF-MS and LC/MS-MS and identified thirty-five compounds. Other than the abundant chlorogenic acid (*trans*- and *cis*- form of 9) and its isomers, caffeoyl-shikimic acid (0.48% of the total combined area of all compounds), feroylquinic acid isomer (0.83%), and traces of caffeic acid (0.16%) were also quantified; percentages represent the relative share of each compound from the total combined peak area of all detected compounds in the leaves. The second significant group of phenolics is flavonoids. Quercetin-3-*O*-glucuronide is the most abundant flavonol and its concentration ranges from 70% to 93% of total flavonols [66]; other flavonols in bilberry leaves are quercetin-3-*O*-β-galactoside (4.06%), quercetin-3-*O*-(4''-HMG)-α-rhamnoside (3.48%), quercetin-3-*O*-arabinoside (2.92%), quercetin-3-*O*-glucoside (0.99%), quercitrin (0.73%), and quercetin (0.03%), as well as three kaempferol glycosides (almost 1.5%) [19,34,66]. Hokkanen *et al.* [34] have in addition detected several other bioactive compounds in these leaves, such as flavan-3-ols comprising 2.0% of the total combined area of all compounds, six different isomers of cinchonain (18.5%), three proanthocyanidins (1.8%), and two coumaroyl iridoids (almost 1.0%). In another study, powdered leaves were extracted with diethyl ether and analyzed with regard to the triterpenoid content. Even though the triterpenes in the leaves comprised only the 4%–6% of those in the respective fruits, several compounds were identified in notable amounts (4.4–4.7 mg/g of dry leaf weight). The predominant compound was β-amyrin (23), followed by oleanane- and ursane-type triterpenes. The triterpene oleanolic and ursolic acids (22) were also identified [36].

The researchers have shown that the collection time of bilberry leaves greatly determines their phenolic content [66]. Hydroxycinnamic acid content strongly decreases during leaf development, while the content of flavonoids increases rapidly until mid-July, as flavonoids are formed later than phenolic acids in the biosynthetic process [38]. As the foliage ages, the color of bilberry leaves changes from green to red during early autumn; this alteration is attributed to differences in the phytochemical composition. Riihinen *et al.* [35] showed that red bilberry leaves contain anthocyanins, even though in very small concentration (0.882 mg/g frozen sample), in contrast with the green leaves where anthocyanins are absent. In addition, quercetin (14) (10.369 mg/g), kaempferol (15) (0.244 mg/g), *p*-coumaric (7) (6.007 mg/g), caffeic (1), or ferulic (4) acids (16.249 mg/g) are higher in the red bilberry leaves compared with the green ones (3.369, 0.171, 2.989, and 7.808 mg/g, respectively) [64]. On the other hand, both green and red leaves contain proanthocyanidins (red: 0.438 mg/g and green: 0.987 mg/g of frozen sample), especially procyanidin; thus, it has been suggested that these leaves should be viewed as a good source of proanthocyanidin-containing products and could be used in cosmetics and pharmaceuticals similarly to the phenolic compounds of green tea [35].

Bilberry has several traditional uses in folk medicine. Decoctions and infusions of its leaves are used for their diuretic, astringent, and antiseptic properties of the urinary tract. Bilberry leaf aqueous extracts are also useful as antibacterials and against inflammation, especially inflammation of the oral cavity [29]. In addition, a widespread use against diabetes has been reported [29].

Despite its regular and significant use as antidiabetic, *Vaccinium myrtillus* leaves have only been rarely investigated and the results are quite contradictory [39]. In alloxan-diabetic mice, a reduction in blood glucose levels by about 10% was reported in the early 1990s, but unfortunately, these experiments are not documented in detail [41]. Cignarella *et al.* [67] tested a dried hydroalcoholic extract of *V. myrtillus* leaves in streptozotocin-diabetic rats (3.0 g extract per kg body weight) and recorded lipid-lowering activity, *i.e.*, decrease by 39% of the triglycerides in the blood of dyslipidemic animals. In addition, they recorded a 26% decrease of plasma glucose levels, but the effect was characterized as "statistically, though not biologically significant". Petlevski *et al.* [68] tested a multi-ingredient preparation composed of Myrtilli folium and nine other plant extracts, patented as an antidiabetic remedy in Croatia; they found a decrease in blood glucose and fructosamine levels in alloxan-induced non-obese diabetic mice. In studies where *Vaccinium* species were introduced in screening programs that aimed at identifying alpha-amylase inhibitors and activators of the human peroxisome proliferator-activated receptor gamma, bilberry leaf extracts showed some activity in both models [69] indicating possible antidiabetic properties. Finally, cinhonains might play significant role in the blood glucose lowering effect as they have been found to induce insulin secretion in both *in vitro* and *in vivo* experiments in rats [67].

Bilberry leaves have been investigated for their antistaphylococcal activity; significant effectiveness against *S. aureus* enhancing, at the same time, the bactericidal potential of vancomycin and linezolid in combination, which has been documented [32]. Finally, bilberry leaves have been explored for their protective activities against cancer. Flavonoids, caffeic acid, and chlorogenic acid were isolated from Sakhalin bilberry *Vaccinium smallii* leaves and were studied as cancer-preventive agents; they inhibited epidermal growth factor (EGF)-induced neoplastic transformation of mouse cells, without exerting any toxic effects [33].

6. Blueberry (*Vaccinium* sp.) Leaves

The term blueberries describes several different taxa of the genus *Vaccinium*; rabbiteye (*V. virgatum*), northern highbush (*V. corymbosum*), southern highbush (*V. formosum*), and lowbush (*V. angustifolium*) blueberries are the commonest.

Red dried leaves of *V. corymbosum* from Drama (region of Macedonia, Greece) were used for the preparation of a decoction (crude extract), which was further fractionated with the organic solvents ethyl acetate and butanol in our laboratory [70]. Analysis was performed by LC-ESI/MS and HPLC-DAD, and twenty different compounds were identified, mainly phenolic acids and flavonols. Interestingly, these two groups were in almost equal concentration in the crude extract (69.34 mg chlorogenic acid equivalents/g dry extract and 67.48 mg quercetin-3-*O*-galactoside equivalents/g dry extract, respectively) [70]; as in bilberry leaves, the most abundant compound was chlorogenic acid (9) (61.31 mg/g dry extract). LC-MS analysis showed the presence of quinic and caffeic acid (1), four myricetin glycosides, one kaempferol rutinoside, and seven quercetin glycosides, as well as quercetin aglycone (14). Hyperoside, isoquercetin, and rutin were the principle flavonoids (12.09 mg/g, 4.60 mg/g, and 3.16 mg/g dry extract). Moreover, we have detected proanthocyanidin B1/B2, kandelin and cinhonain. Kandelin was also reported in the *Vaccinium ashei* leaves [42], while cinhonains have been identified in bilberry leaves [34]. The absence of anthocyanins from the decoction was notable; however, it could also be attributed to the method of the extraction [70].

Wang *et al.* [40] published a study where 104 different cultivars of blueberries (rabbiteye, northern highbush, and southern highbush) were examined with respect to their phytochemical composition and antioxidant properties. Using HPLC–ESI–MS^2 analysis, they identified three anthocyanins (cyanidin 3-*O*-glucoside, cyanidin 3-*O*-glucuronide, cyanidin 3-*O*- arabinoside) in the blueberry leaf methanolic extracts, even though in different quantities; northern highbush blueberries showed the higher total anthocyanin content (TAC). Nevertheless, TAC, which was measured semiquantitatively by linear regression of commercial standards, was almost ten times lower than that of the respective fruit in each cultivar, ranging from 0.09 to 4.4 mg cyanidin 3-*O*-glucoside equivalents/g dry weight. Leaf anthocyanins were not detected in some cultivars. Moreover, they detected four different proanthocyanidins but in very small amounts in the highbush blueberries (0.36–8.38 mg rutin equivalents/g dry weight) [40].

Leaf tissue maturation plays a significant role in the phytochemical composition of this species. Riihinen *et al.* [35] have showed that the red leaves of *V. corymbosum* contain higher amounts of quercetin (14) (3.530 mg/g frozen sample) and kaempferol (15) (0.505 mg/g), as well as of *p*-coumaric (7) (3.060 mg/g), caffeic (1) or ferulic (4) acids (19.870 mg/g) than the green leaves (1.784, 0.191, 0.490, 7.537 mg/g frozen sample, respectively). Solar radiation increases the content of the above-mentioned flavonols and hydroxycinnamic acids, probably due to their role in photo-protection [37]. This explains the higher content of those compounds in the red leaves compared to the green leaves. In addition, red leaves contain a very small amount of anthocyanins, which are absent from the green. On the other hand, prodelphinidins and procyanidins are present almost in the same quantity in both types of leaves (red: 0.485 mg/g, green: 0.468 mg/g frozen sample) [35].

Harris *et al.* [43] investigated the phytochemical profile of *V. angustifolium* leaves, demonstrating its high similarity with the highbush blueberry leaves. They identified ten different compounds in an ethanolic leaf extract. Chlorogenic acid (9) was the most abundant phenolic; it was 30 times more concentrated in the leaf extract (31.19 mg/g dry matter) than in the respective fruit (1.54 mg/g dry matter) and over 100 times more concentrated than in the respective stem or root extracts (0.09 and 0.03 mg/g dry matter, respectively). Moreover, they detected in significant amounts the flavan-3-ols epicatechin (20) and catechin (19) and in ratio roughly 1:1; they also quantified four quercetin glycosides (in total 9.65 mg/g dry matter), as well as quercetin aglycone (1.24 mg/g dry matter) (13). Quercetin-3-*O*-glucoside and quercetin-3-*O*-arabinoside accounted for 36% and 28%, respectively, of the quantified quercetin glycosides. Caffeic acid (1) was found in traces (0.36 mg/g dry matter); chlorogenic acid isomers (10, 11), quercetin-hexoside, quercetin-pentoside, and rutin were detected, but not quantified. Exactly the same compounds were quantified in another ethanolic extract of lowbush blueberry leaves, but in this case the measured quantities were almost three-fold higher than in the study of Harris *et al.* [44]. Anthocyanins were not detected in any of these studies.

Various members of the *Vaccinium* genus, other than bilberry, such as *Vaccinium macrocarpon* and *Vaccinium angustifolium*, are reputed to possess antidiabetic activity [71] and have been used extensively as traditional medicines for the treatment of diabetic symptoms [72]. Martineau *et al.* [44] demonstrated the significant antidiabetic activity of lowbush blueberry leaves *in vitro* with various cell-based bioassays. However, despite the widespread traditional use against diabetes, screening of current literature revealed the absence of investigations other than that of Martineau *et al.* [44].

The majority of studies focus on the antioxidant activities of blueberry leaves, which are related to their rich in phenolics composition [47]. In line with these findings, we have also demonstrated the high antioxidant capacity of *V. corymbosum* leaf decoction and its capability to bind iron ions [70]. In addition to ferrous chelation activity, in several *in vitro* experiments we have proven that quercetin (i) is totally oxidized by selenite ions, and (ii) it chelates calcium ions probably via the hydroxyl groups of A and B rings of the flavonoids. These observations were related to the protective activity that we have recorded against selenite-induced ocular cataract and selenite-induced oxidative damage in the brain and liver of neonatal rats [45,70]. Finally, highbush blueberry leaf extract acts as an antimicrobial agent, especially against *Salmonella typhymurium* and *Enterococcus faecalis* [46].

7. Cranberry (*Vaccinium* sp.) Leaves

Cranberries are a group of ecergreen dwarf shrubs or trailing vines in the subgenus of oxycoccus of the genus *Vaccinium*. In North America, cranberry may refer to *V. macrocarpon*, *V. microcarpon*, or *V. erythrocarpon*, whereas in Britain, cranberry usually refers to the native *V. oxycoccos*. In a recent comparative study, Teleszko *et al.* [73] analyzed the phytochemical composition of fruits and leaves of several berry species by UPLC-PDA/FL and LC/MS; among them, cranberry leaves were the second richest source of phenolics, richer than bilberry and blackcurrant leaves. The major polyphenolic group was proanthocyanidins, followed by flavonols. Proanthocyanidins (47.18 mg/g dry leaves), flavan-3-ols (27.76 mg/g), phenolic acids (2.36 mg/g), and flavonols (33.64 mg/g) were in higher concentration than in the respective fruits [73]. In addition, Neto *et al.* [74] have performed an HPLC-MS analysis of the phenolic profile of two cultivars, *i.e.*, in *Early Black* and *Howes*; the phenolic acids are mainly chlorogenic and neo-chlorogenic acid, as well as 3-*O*- and 5-*O*-coumaroylquinic acids. The principle flavonols were hyperoside and quercetin-3-*O*-rhamnoside, while quercetin-3-*O*-xyloside and quercetin-3-*O*-arabinoside were also detected. Procyanidin A2 was the identified catechin dimer. Finally, they documented two coumaroyl iridoid isomers (*trans*- and *cis*-form) previously reported in cranberry fruit [75]. All these compounds were in higher content in the cultivar *Early Black*.

The high phenolic content of cranberry leaves seems to be associated with the significant antioxidant potential that has been recorded with different methods [73]. Cranberries, however, are mostly known for their use against urinary tract infections. A randomized, double-blind, placebo-controlled cross-over experimental trial with 12 participants showed that the consumption of a cranberry-leaf beverage increased blood glutathione peroxidase activity, indicating its antioxidant capacity and inhibited the *ex vivo* adhesion of P-fimbriated *Escherichia coli* bacteria in urine, suggesting that cranberry leaf extracts may help to improve urinary tract health [76].

8. Lingonberry (*Vaccinium vitis-idaea*) Leaves

The green leaves of lingonberry (*Vaccinium vitis-idaea*) have similar phytochemical profile with those of bilberry [34,66]. Ieri *et al.* [66] and Hokkanen *et al.* [34] have quantified a great number of phenolics in the methanolic and hydro-alcoholic leaf extracts of lingonberry, respectively. In general, hydroxycinnamic acids and flavonols were the most abundant compounds. In the methanolic extract, flavonol content was higher than hydroxycinnamic acids, but in the hydro-alcoholic extract the opposite was observed, as expected. In both cases, the main acid was 2-*O*-caffeoylarbutin (12) (14%–35% of total phenols), which is not present in other berry leaves. Other phenolic acids detected in the methanolic extract were chlorogenic acid (3.55% of the total combined peak area of all compounds), coumaroyl quinic acid isomers (3.81%), caffeic acid (0.61%), *p*-coumaric acid (0.64%), and caffeoyl-shikimic acid (0.18%) [34]. However, lingonberry leaves contain coumaroyl- and caffeoyl-hexose hydroxyphenols (1.85% and 1.03% of the total combined peak area of all compounds) which are not present in bilberry leaves.

With respect to the flavonols, the principle compound was quercetin-3-*O*-(4″-HMG)-α-rhamnoside in both studies comprising 5%–6% of total phenols in the hydroalcoholic extract and 32% of the total combined peak area of all compounds in the methanolic extract. Rutin (7.59% of the total combined peak area of all compounds), hyperoside (6.30%), and quercetrin (5.37%) were also detected in significant amounts in the methanolic extract, while traces of four more quercetin glycosides were also quantified. Furthermore, very small concentrations of kaempferol glycosides (8.03%), proanthocyanidins (1.4%), and coumaroyl iridoids (7.4%) were found [34].

Lingonberry leaves have similar traditional uses in folk medicine with the bilberry leaves [66]. They were mainly used as diuretics as well as for their antiseptic activity in urinary tract, probably due to the high content of tannins, especially arbutin and its derivatives [66]. Recently, the ethanolic extract of lingonberry leaves has shown significant antitussive, anti-inflammatory, and anti-catarrhal properties in rats [77]. In addition, Vyas *et al.* [78] has demonstrated that the acetone extract of these leaves possesses significant neuroprotective effect *in vitro* against glutamate-mediated excitotoxicity.

9. Conclusions

Despite their traditional uses, berry leaves are seldom used nowadays, in contrast to berry fruits, which are considered foods with significant health benefits. However, recent investigations have revealed that the traditional therapeutic properties of berry leaves may be valid. Moreover, the study of the phytochemical composition of berry leaves points out that they can be viewed as rich sources of bioactive natural products, e.g., tannins in raspberry and bilberry leaves, and chlorogenic acid in blueberry leaves, whereas other berry leaves, such as lingonberry, contain unique phenolics like arbutins. The phenolic compounds of the leaves are known antioxidant and anti-inflammatory agents (quercetin and kaempferol derivatives), and have hypoglycemic (cinchonains) and antimicrobial (ellagitannins) properties. Several studies and reviews have pointed out the anti-inflammatory activities of naturally-occurring compounds; the most effective are usually the aglycon forms of flavonoids (quercetin, kaempferol) and most of their actions are related to their ability to inhibit cytokine, chemokine release, and to be implicated in the molecular paths of the synthesis and/or action of adhesion molecules [79,80].

Epidemiological and meta-analyses studies suggest an inverse relationship between flavonoid-rich diets and development of many aging-associated diseases including cancers, cardiovascular disease, diabetes, osteoporosis, and neurodegenerative disorders [76]. Dietary flavonoids exert their anti-diabetic effects by targeting various cellular signaling pathways in pancreas, liver, and skeletal muscle; by influencing β-cell mass and function, as well as energy metabolism and insulin sensitivity in peripheral tissues [81]. Even though scientific literature specifically on the effectiveness of berry leaf consumption is extremely limited, the beneficial properties of individual flavonoids *in vitro* hold promise of positive outcomes. Nevertheless, *in vivo* studies with berry leaf extracts to evaluate modification of various biomarkers of disease and potential toxicity are needed. Additionally, bioavailability and pharmacokinetic studies in healthy human subjects, as well as carefully-designed and targeted intervention trials that would evaluate the impact of berry leaf-derived products on the prevention or the progress of specific disorders, are necessary. The forgotten berry leaves have just been "re-discovered" and may be viewed as sources of valuable bioactive compounds with health-promoting and disease-preventing properties.

Conflicts of Interest: The authors declare no conflict of interest.

Abbreviations

The following abbreviations are used in this manuscript:

DAD   Diode Array Detection
EGF   Epidermal Growth Factor
EMA   European Medicines Agency
ESI   Electrospray ionization
HMG   Hydroxymethylglutaric acid
HPLC   High Performance Liquid Chromatography
HPTLC High Performance Thin Layer Chromatography

IR	Infrared spectroscopy
LC	Liquid Chromatography
MS	Mass Spectrometry
NMR	Nuclear Magnetic Resonance
TOF	Time-of-flight

References

1. European Commission. *Functional Foods*; Publications Office of the European Union: Luxembourg, 2010.
2. Beattie, J.; Crozier, A.; Duthie, G.G. Potential health benefits of berries. *Curr. Nutr. Food Sci.* **2004**, *1*, 71–86. [CrossRef]
3. Mattila, P.; Hellstrom, J.; Torronen, R. Phenolic acids in berries, fruits, and beverages. *J. Agric. Food Chem.* **2006**, *54*, 7193–7199. [CrossRef] [PubMed]
4. Maatta-Riihinen, K.R.; Kamal-Eldin, A.; Mattila, P.H.; Gonzalez-Paramas, A.M.; Torronen, A.R. Distribution and contents of phenolic compounds in eighteen Scandinavian berry species. *J. Agric. Food Chem.* **2004**, *52*, 4477–4486. [CrossRef] [PubMed]
5. Maatta-Riihinen, K.R.; Kamal-Eldin, A.; Torronen, A.R. Identification and quantification of phenolic compounds in berries of Fragaria and Rubus species (family Rosaceae). *J. Agric. Food Chem.* **2004**, *52*, 6178–6187. [CrossRef] [PubMed]
6. Kylli, P. Berry Phenolics: Isolation, Analysis, Identification and Antioxidant Properties. Ph.D. Thesis, University of Helsinki, Helsinki, Finland, 2011.
7. Häkkinen, S.; Heinonen, M.; Kärenlampi, S.; Mykkänen, H.; Ruuskanen, J.; Törrönen, R. Screening of selected flavonoids and phenolic acids in 19 berries. *Food Res. Int.* **1999**, *32*, 345–353. [CrossRef]
8. Taruscio, T.G.; Barney, D.L.; Exon, J. Content and profile of flavanoid and phenolic acid compounds in conjunction with the antioxidant capacity for a variety of northwest Vaccinium berries. *J. Agric. Food Chem.* **2004**, *52*, 3169–3176. [CrossRef] [PubMed]
9. Kahkonen, M.P.; Heinonen, M. Antioxidant activity of anthocyanins and their aglycons. *J. Agric. Food Chem.* **2003**, *51*, 628–633. [CrossRef] [PubMed]
10. Koponen, J.M.; Happonen, A.M.; Mattila, P.H.; Torronen, A.R. Contents of anthocyanins and ellagitannins in selected foods consumed in Finland. *J. Agric. Food Chem.* **2007**, *55*, 1612–1619. [CrossRef] [PubMed]
11. Gu, L.; Kelm, M.A.; Hammerstone, J.F.; Beecher, G.; Holden, J.; Haytowitz, D.; Gebhardt, S.; Prior, R.L. Concentrations of proanthocyanidins in common foods and estimations of normal consumption. *J. Nutr.* **2004**, *134*, 613–617. [PubMed]
12. Committee on Herbal Medicinal Products (HMPC). Call for Scientific Data for Use in HMPC Assessment Work on *Fragaria vesca* L., Folium. European Medicines Agency: London, UK, 2014.
13. Committee on Herbal Medicinal Products (HMPC). Community Herbal Monograph on *Rubus idaeus* L., Folium. European Medicines Agency: London, UK, 2012.
14. Committee on Herbal Medicinal Products (HMPC). Community Herbal Monograph on *Ribes nigrum* L., Folium. European Medicines Agency: London, UK, 2009.
15. Committee on Herbal Medicinal Products (HMPC). Community Herbal Monograph on *Arctostaphylos uva-ursi* (L.) Spreng., Folium. European Medicines Agency: London, UK, 2009.
16. Connolly, T.J. Newberry crater: A ten-thousand-year record of human occupation and environmental change in the basin-plateau borderlands. *Plains Anthropol.* **2003**, *48*, 168–170.
17. Riaz, M.A.; Rahman, N.U. Antimicrobial screening of fruit, leaves, root and stem of *Rubus fruticosus* L. *J. Med. Plants Res.* **2011**, *5*, 5920–5924.
18. Sarkar, A.; Ganguly, S.N. Rubitic acid, a new triterpene acid from *Rubus fruticosus*. *Phytochemistry* **1978**, *17*, 1983–1985. [CrossRef]
19. Oszmianski, J.; Wojdylo, A.; Gorzelany, J.; Kapusta, I. Identification and characterization of low molecular weight polyphenols in berry leaf extracts by HPLC-DAD and LC-ESI/MS. *J. Agric. Food Chem.* **2011**, *59*, 12830–12835. [CrossRef] [PubMed]
20. Raudsepp, P.; Kaldmae, H.; Kikas, A.; Libek, A.V.; Püssa, T. Nutritional quality of berries and bioactive compounds in the leaves of black currant (*Ribes nigrum* L.) cultivars evaluated in Estonia. *J. Berry Res.* **2010**, *1*, 53–59.

21. Committee on Herbal Medicinal Products (HMPC). Assessment Report on *Rubus idaeus* L., Folium. European Medicines Agency: London, UK, 2012.

22. Oszmiański, J.; Wojdyło, A.; Nowicka, P.; Teleszko, M.; Cebulak, T.; Wolanin, M. Determination of phenolic compounds and antioxidant activity in leaves from wild *Rubus* L. species. *Molecules* **2015**, *20*, 4951–4966. [CrossRef] [PubMed]

23. Abu-Shandi, K.; Al-Rawashdeh, A.; Al-Mazaideh, G.; Abu-Nameh, E.; Al-Amro, A.; Al-Soufi, H.; Al-Ma'abreh, A.; Al-Dawdeyah, A. A novel strategy for the identification of the medicinal natural products in *Rubus fruticosus* plant by using GC/MS technique: A study on leaves, stems and roots of the plant. *Adv. Anal. Chem.* **2015**, *5*, 31–41.

24. Gudej, J.; Tomczyk, M. Determination of flavonoids, tannins and ellagic acid in leaves from *Rubus* L. species. *Arch. Pharm. Res.* **2004**, *27*, 1114–1119. [CrossRef] [PubMed]

25. Zia-Ul-Haq, M.; Riaz, M.; de Feo, V.; Jaafar, H.; Moga, M. *Rubus fruticosus* L.: Constituents, biological activities and health related uses. *Molecules* **2014**, *19*, 10998–11029. [CrossRef] [PubMed]

26. Robinson, G.M.; Robinson, R. A survey of anthocyanins. *Biochem. J.* **1931**, *25*, 1687–1705. [CrossRef] [PubMed]

27. Mukherjee, M.; Ghatak, K.L.; Ganguly, S.N.; Antoulas, S. Rubinic acid, a triterpene acid from *Rubus fruticosus*. *Phytochemistry* **1984**, *23*, 2581–2582. [CrossRef]

28. Greenway, F.L.; Zhijun, L.; Woltering, E.A. Angiogenic Agents from Plant Extracts, Gallic Acid and Derivatives. U.S. Patent 20,070,031,332, 8 February 2007.

29. Wang, S.Y.; Lin, H.S. Antioxidant activity in fruits and leaves of blackberry, raspberry, and strawberry varies with cultivar and developmental stage. *J. Agric. Food Chem.* **2000**, *48*, 140–146. [CrossRef] [PubMed]

30. Scientific Committee of the British Herbal Medicine Association. Rubus. In *British Herbal Pharmacopoeia*, 2nd Consolidated; British Herbal Medicine Association: Bournemouth, UK, 1983; pp. 181–182.

31. Jan, G. Kaemferol and quercetin glycosides from *Rubus idaeus* L. leaves. *Acta Polon. Pharm.* **2003**, *60*, 313–316.

32. Sadowska, B.; Paszkiewicz, M.; Podsedek, A.; Redzynia, M.; Rozalska, B. *Vaccinium myrtillus* leaves and *Frangula alnus* bark derived extracts as potential antistaphylococcal agents. *Acta Biochim. Pol.* **2014**, *61*, 163–169. [PubMed]

33. Mechikova, G.Y.; Kuzmich, A.S.; Ponomarenko, L.P.; Kalinovsky, A.I.; Stepanova, T.A.; Fedorov, S.N.; Stonik, V.A. Cancer-preventive activities of secondary metabolites from leaves of the bilberry *Vaccinium smallii* A. Gray. *Phytother. Res.* **2010**, *24*, 1730–1732. [CrossRef] [PubMed]

34. Hokkanen, J.; Mattila, S.; Jaakola, L.; Pirttila, A.M.; Tolonen, A. Identification of phenolic compounds from lingonberry (*Vaccinium vitis-idaea* L.), bilberry (*Vaccinium myrtillus* L.) and hybrid bilberry (*Vaccinium x intermedium* Ruthe L.) leaves. *J. Agric. Food Chem.* **2009**, *57*, 9437–9447. [CrossRef] [PubMed]

35. Riihinen, K.; Jaakola, L.; Karenlampi, S.; Hohtola, A. Organ-specific distribution of phenolic compounds in bilberry (*Vaccinium myrtillus*) and "northblue" blueberry (*Vaccinium corymbosum x V. angustifolium*). *Food Chem.* **2008**, *110*, 156–160. [CrossRef] [PubMed]

36. Szakiel, A.; Paczkowski, C.; Huttunen, S. Triterpenoid content of berries and leaves of bilberry *Vaccinium myrtillus* from Finland and Poland. *J. Agric. Food Chem.* **2012**, *60*, 11839–11849. [CrossRef] [PubMed]

37. Jaakola, L.; Maatta-Riihinen, K.; Karenlampi, S.; Hohtola, A. Activation of flavonoid biosynthesis by solar radiation in bilberry (*Vaccinium myrtillus* L.) leaves. *Planta* **2004**, *218*, 721–728. [PubMed]

38. Martz, F.; Jaakola, L.; Julkunen-Tiitto, R.; Stark, S. Phenolic composition and antioxidant capacity of bilberry (*Vaccinium myrtillus*) leaves in Northern Europe following foliar development and along environmental gradients. *J. Chem. Ecol.* **2010**, *36*, 1017–1028. [CrossRef] [PubMed]

39. Helmstadter, A.; S chuster, N. *Vaccinium myrtillus* as an antidiabetic medicinal plant—Research through the ages. *Parmazie* **2010**, *65*, 315–321.

40. Wang, L.J.; Wu, J.; Wang, H.X.; Li, S.S.; Zheng, X.C.; Du, H.; Xu, Y.J.; Wang, L.S. Composition of phenolic compounds and antioxidant activity in the leaves of blueberry cultivars. *J. Funct. Foods* **2015**, *16*, 295–304. [CrossRef]

41. Hänsel, R.; Keller, K.; Rimpler, H.; Schneider, G. Drogen p–z. In *Hager's Handbuch der Pharmazeutischen Praxis*, 5th ed.; Springer-Verlag GmbH: Berlin Heidelberg, Germany, 1996; pp. 1051–1067.

42. Matsuo, Y.; Fujita, Y.; Ohnishi, S.; Tanaka, T.; Hirabaru, H.; Kai, T.; Sakaida, H.; Nishizono, S.; Kouno, I. Chemical constituents of the leaves of rabbiteye blueberry (*Vaccinium ashei*) and characterisation of polymeric proanthocyanidins containing phenylpropanoid units and a-type linkages. *Food Chem.* **2010**, *121*, 1073–1079.

43. Harris, C.S.; Burt, A.J.; Saleem, A.; Le, P.M.; Martineau, L.C.; Haddad, P.S.; Bennett, S.A.; Arnason, J.T. A single HPLC-PAD-APCI/MS method for the quantitative comparison of phenolic compounds found in leaf, stem, root and fruit extracts of *Vaccinium angustifolium*. *Phytochem. Anal.* **2007**, *18*, 161–169. [CrossRef] [PubMed]

44. Martineau, L.C.; Couture, A.; Spoor, D.; Benhaddou-Andaloussi, A.; Harris, C.; Meddah, B.; Leduc, C.; Burt, A.; Vuong, T.; Mai Le, P.; *et al.* Anti-diabetic properties of the Canadian lowbush blueberry *Vaccinium angustifolium* Ait. *Phytomedicine* **2006**, *13*, 612–623. [PubMed]

45. Ferlemi, A.V.; Makri, O.E.; Mermigki, P.G.; Lamari, F.N.; Georgakopoulos, C.D. Quercetin glycosides and chlorogenic acid in highbush blueberry leaf decoction prevent cataractogenesis *in vivo* and *in vitro*: Investigation of the effect on calpains, antioxidant and metal chelating properties. *Exp. Eye Res.* **2016**, *145*, 258–268. [CrossRef] [PubMed]

46. Pervin, M.; Hasnat, M.A.; Lim, B.O. Antibacterial and antioxidant activities of *Vaccinium corymbosum* L. leaf extract. *Asian Pac. J. Trop. Dis.* **2013**, *3*, 444–453. [CrossRef]

47. Piljac-Zegarac, J.; Belscak, A.; Piljac, A. Antioxidant capacity and polyphenolic content of blueberry (*Vaccinium corymbosum* L.) leaf infusions. *J. Med. Food* **2009**, *12*, 608–614. [CrossRef] [PubMed]

48. Pharmacopée Française. *Cassis (feuille de) Ribis nigri Folium*; Ministere de la Sante: Paris, France, 1996.

49. Trajkovski, V. Resistance to *Sphaerotheca mors-uvae* (Schw.) Berk. in *Ribes nigrum* L. *Swed. J. Agric. Res.* **1974**, *4*, 99–108.

50. Vagiri, M.; Conner, S.; Stewart, D.; Andersson, S.C.; Verrall, S.; Johansson, E.; Rumpunen, K. Phenolic compounds in blackcurrant (*Ribes nigrum* L.) leaves relative to leaf position and harvest date. *Food Chem* **2015**, *172*, 135–142. [CrossRef] [PubMed]

51. Vagiri, M.; Ekholm, A.; Andersson, S.C.; Johansson, E.; Rumpunen, K. An optimized method for analysis of phenolic compounds in buds, leaves, and fruits of black currant (*Ribes nigrum* L.). *J. Agric. Food Chem.* **2012**, *60*, 10501–10510. [CrossRef] [PubMed]

52. Tits, M.; Poukens, P.; Angenot, L.; Dierckxsens, Y. Thin-layer chromatographic analysis of proanthocyanidins from *Ribes nigrum* leaves. *J. Pharm. Biomed. Anal.* **1992**, *10*, 1097–1100. [CrossRef]

53. Tits, M.; Angenot, L.; Poukens, P.; Warin, R.; Dierckxsens, Y. Prodelphinidins from *Ribes nigrum*. *Phytochemistry* **1992**, *31*, 971–973. [CrossRef]

54. Committee on Herbal Medicinal Products (HMPC). *Assessment Report on Ribes nigrum L., Folium*; European Medicines Agency: London, UK, 2010.

55. Andersson, J.; Bosvik, R.; Von Sydow, E. The composition of the essential oil of black currant leaves (*Ribes nigrum* L.). *J. Sci. Food Agric.* **1963**, *14*, 834–840. [CrossRef]

56. Tabart, J.; Franck, T.; Kevers, C.; Pincemail, J.; Serteyn, D.; Defraigne, J.O.; Dommes, J. Antioxidant and anti-inflammatory activities of *Ribes nigrum* extracts. *Food Chem.* **2012**, *131*, 1116–1122. [CrossRef]

57. Grayer, R.J.; Kokubun, T. Plant–fungal interactions: The search for phytoalexins and other antifungal compounds from higher plants. *Phytochemistry* **2001**, *56*, 253–263. [CrossRef]

58. Xu, Y.; Zhang, Y.; Chen, M. Effective Fractions of *Rubus fruticosus* Leaf, Its Pharmaceutical Composition and Uses for Prevention and Treatment of Diabetes. China Patent CN1788755 A, 21 June 2006.

59. Alonso, R.; Cadavid, I.; Calleja, J.M. A preliminary study of hypoglycemic activity of *Rubus fruticosus*. *Planta Med.* **1980**, *40*, 102–106. [CrossRef] [PubMed]

60. Osório e Castro, V.R. Chromium and zinc in a series of plants used in Portugal in the herbal treatment of non-insulinized diabetes. *Acta Aliment.* **2001**, *30*, 333–342. [CrossRef]

61. Durgo, K.; Belscak-Cvitanovic, A.; Stancic, A.; Franekic, J.; Komes, D. The bioactive potential of red raspberry (*Rubus idaeus* L.) leaves in exhibiting cytotoxic and cytoprotective activity on human laryngeal carcinoma and colon adenocarcinoma. *J. Med. Food* **2012**, *15*, 258–268. [CrossRef] [PubMed]

62. Brandely, P. Raspberry leaf. In *British Herbal Compendium. A Handbook of Scientific Information on Widely Used Plant Drugs*; British Herbal Medicine Association: Bournemouth, UK, 2006; pp. 328–331.

63. Parsons, M.; Simpson, M.; Ponton, T. Raspberry leaf and its effect on labour: Safety and efficacy. *Aust. Coll. Midwives Inc. J.* **1999**, *12*, 20–25. [CrossRef]

64. Jing, Z.; Pistilli, M.J.; Holloway, A.C.; Crankshaw, D.J. The effects of commercial preparations of red raspberry leaf on the contractility of the rat's uterus *in vitro*. *Reprod. Sci.* **2010**, *17*, 494–501. [CrossRef] [PubMed]

65. Simpson, M. Raspberry leaf in pregnancy: Its safety and efficacy in labor. *J. Midwifery Women's Health* **2001**, *46*, 51–59. [CrossRef]

66. Ieri, F.; Martini, S.; Innocenti, M.; Mulinacci, N. Phenolic distribution in liquid preparations of *Vaccinium myrtillus* L. and *Vaccinium vitis idaea* L. *Phytochem. Anal.* **2013**, *24*, 467–475. [CrossRef] [PubMed]

67. Cignarella, A.; Nastasi, M.; Cavalli, E.; Puglisi, L. Novel lipid-lowering properties of *Vaccinium myrtillus* L. leaves, a traditional antidiabetic treatment, in several models of rat dyslipidaemia: A comparison with ciprofibrate. *Thromb. Res.* **1996**, *84*, 311–322. [CrossRef]

68. Petlevski, R.; Hadžija, M.; Slijepčević, M.; Juretić, D. Effect of "antidiabetis" herbal preparation on serum glucose and fructosamine in NOD mice. *J. Ethnopharmacol.* **2001**, *75*, 181–184. [CrossRef]

69. Rau, O.; Wurglics, M.; Dingermann, T.; Abdel-Tawab, M.; Schubert-Zsilavecz, M. Screening of herbal extracts for activation of the human peroxisome proliferator-activated receptor. *Pharmazie* **2006**, *61*, 952–956. [PubMed]

70. Ferlemi, A.V.; Mermigki, P.G.; Makri, O.E.; Anagnostopoulos, D.; Koulakiotis, N.S.; Margarity, M.; Tsarbopoulos, A.; Georgakopoulos, C.D.; Lamari, F.N. Cerebral area differential redox response of neonatal rats to selenite-induced oxidative stress and to concurrent administration of highbush blueberry leaf polyphenols. *Neurochem. Res.* **2015**, *40*, 2280–2292. [CrossRef] [PubMed]

71. Chambers, B.K.; Camire, M.E. Can cranberry supplementation benefit adults with type 2 diabetes? *Diabetes Care* **2003**, *26*, 2695–2696. [CrossRef] [PubMed]

72. Jellin, J.M.; Gregory, P.J.; Batz, F.; Hitchens, K. Natural medicines comprehensive database. *Pharm. Lett./Prescr. Lett.* **2005**, 2239.

73. Teleszko, M.; Wojdyło, A. Comparison of phenolic compounds and antioxidant potential between selected edible fruits and their leaves. *J. Funct. Foods* **2015**, *14*, 736–746. [CrossRef]

74. Neto, C.C.; Salvas, M.R.; Autio, W.R.; van den Heuvel, J.E. Variation in concentration of phenolic acid derivatives and quercetin glycosides in foliage of cranberry that may play a role in pest deterrence. *J. Am. Soc. Hortic. Sci.* **2010**, *135*, 494–500.

75. Turner, A.; Chen, S.N.; Nikolic, D.; van Breemen, R.; Farnsworth, N.R.; Pauli, G.F. Coumaroyl iridoids and a depside from cranberry (*Vaccinium macrocarpon*). *J. Nat. Prod.* **2007**, *70*, 253–258. [CrossRef] [PubMed]

76. Mathison, B.D.; Kimble, L.L.; Kaspar, K.L.; Khoo, C.; Chew, B.P. Consumption of cranberry beverage improved endogenous antioxidant status and protected against bacteria adhesion in healthy humans: A randomized controlled trial. *Nutr. Res.* **2014**, *34*, 420–427. [CrossRef] [PubMed]

77. Wang, X.; Sun, H.; Fan, Y.; Li, L.; Makino, T.; Kano, Y. Analysis and bioactive evaluation of the compounds absorbed into blood after oral administration of the extracts of *Vaccinium* vitis-idaea in rat. *Biol. Pharm. Bull.* **2005**, *28*, 1106–1108. [CrossRef] [PubMed]

78. Vyas, P.; Kalidindi, S.; Chibrikova, L.; Igamberdiev, A.U.; Weber, J.T. Chemical analysis and effect of blueberry and lingonberry fruits and leaves against glutamate-mediated excitotoxicity. *J. Agric. Food Chem.* **2013**, *61*, 7769–7776. [CrossRef] [PubMed]

79. Hämäläinen, M.; Nieminen, R.; Vuorela, P.; Heinonen, M.; Moilanen, E. Anti-inflammatory effects of flavonoids: Genistein, kaempferol, quercetin, and daidzein inhibit STAT-1 and NF-κB activations, whereas flavone, isorhamnetin, naringenin, and pelargonidin inhibit only NF-κB activation along with their inhibitory effect on iNOS expression and NO production in activated macrophages. *Med. Inflamm.* **2007**, *2007*, 1–10.

80. Calixto, J.B.; Campos, M.M.; Otuki, M.F.; Santos, A.R. Anti-inflammatory compounds of plant origin. Part II. Modulation of pro-inflammatory cytokines, chemokines and adhesion molecules. *Planta Med.* **2004**, *70*, 93–103. [PubMed]

81. Babu, P.V.; Liu, D.; Gilbert, E.R. Recent advances in understanding the anti-diabetic actions of dietary flavonoids. *J. Nutr. Biochem.* **2013**, *24*, 1777–1789. [CrossRef] [PubMed]

antioxidants

MDPI

Article

Anthocyanin Accumulation in Muscadine Berry Skins Is Influenced by the Expression of the MYB Transcription Factors, *MybA1*, and *MYBCS1*

Lillian Oglesby [1], Anthony Ananga [1,2,*], James Obuya [1], Joel Ochieng [3], Ernst Cebert [4] and Violeta Tsolova [1]

[1] Center for Viticulture and Small Fruit Research, College of Agriculture and Food Science, Florida A & M University, 6505 Mahan Drive, Tallahassee, FL 32317, USA; lillian.oglesby@gmail.com (L.O.); jamesobuya@yahoo.com (J.O.); violeta.tsolova@famu.edu (V.T.)
[2] Food Science Program, College of Agriculture and Food Sciences, Florida A & M University, Tallahassee, FL 32307, USA
[3] Faculties of Agriculture and Veterinary Medicine, University of Nairobi, P.O. Box 29053, Nairobi 00625, Kenya; jochieng@uonbi.ac.ke
[4] Department of Biological and Environmental Sciences, Alabama A & M University, 4900 Meridian Street, Normal, AL 35762, USA; ecebert@gmail.com
* Correspondence: anthony.ananga@gmail.com; Tel.: +1-850-412-7393; Fax: +1-850-561-2617

Academic Editor: Dorothy Klimis-Zacas
Received: 16 April 2016; Accepted: 18 September 2016; Published: 12 October 2016

Abstract: The skin color of grape berry is very important in the wine industry. The red color results from the synthesis and accumulation of anthocyanins, which is regulated by transcription factors belonging to the MYB family. The transcription factors that activate the anthocyanin biosynthetic genes have been isolated in model plants. However, the genetic basis of color variation is species-specific and its understanding is relevant in many crop species. This study reports the isolation of *MybA1*, and *MYBCS-1* genes from muscadine grapes for the first time. They are designated as *VrMybA1* (GenBank Accession No. KJ513437), and *VrMYBCS1* (*VrMYB5a*) (GenBank Accession No. KJ513438). The findings in this study indicate that, the deduced *VrMybA1* and *VrMYBCS1* protein structures share extensive sequence similarity with previously characterized plant MYBs, while phylogenetic analysis confirms that they are members of the plant MYB super-family. The expressions of *MybA1*, and *MYBCS1* (*VrMYB5a*) gene sequences were investigated by quantitative real-time PCR using in vitro cell cultures, and berry skin samples at different developmental stages. Results showed that *MybA1*, and *MYBCS1* genes were up-regulated in the veràison and physiologically mature red berry skins during fruit development, as well as in in vitro red cell cultures. This study also found that in ripening berries, the transcription of *VrMybA1*, and *VrMYBCS1* in the berry skin was positively correlated with anthocyanin accumulation. Therefore, the upregulation of *VrMybA1*, and *VrMYBCS1* results in the accumulation and regulation of anthocyanin biosynthesis in berry development of muscadine grapes. This work greatly enhances the understanding of anthocyanin biosynthesis in muscadine grapes and will facilitate future genetic modification of the antioxidants in *V. rotundifolia*.

Keywords: anthocyanins; antioxidants; MYB gene; pigments; muscadine grapes; *MybA1*; *MYBCS1*

1. Introduction

Muscadine grapes (*Vitis rotundifolia* Michx.) are considered important *Vitis* species because they contain several unique flavonoid compounds with beneficial nutraceutical properties [1–3]. They are the only grapes containing ellagic acid in the skin and possess high antioxidant levels in comparison to other fruits [4,5] and contain significantly higher concentration of anthocyanin and phenolic acid [6,7]. Muscadine grapes are also highly resistant to diseases, which enable viticulturists to grow them with minimal applications of pesticides in regions with high disease pressure [8,9].

Several studies have reported that, the ripening of muscadine berry involves several changes in cell wall composition that leads to fruit softening [10–12]. Deytieux et al. and Deytieux-Belleau et al. [13,14] considered these changes to be key phases in growth that can be used to determine the quality of both wine and table grapes, because they immediately precede harvesting. Grape berry is classified as a non-climacteric fruit based on its respiration rates, and several studies have suggested that abscisic acid may play a role in the ripening process, because its increase in concentration correlates with the ripening of berries [15–17]. Since the biosynthetic pathway of flavonoids in muscadine grapes, plays an important role in berry development and ripening, greater understanding of the genes that control such expression is essential. Therefore, investigation of structural and regulatory enzymes that control the flavonoid biosynthesis pathway in muscadine grapes needs to be studied.

During grape berry development, the onset of maturation begins at véraison, that is, the beginning of skin color-change in red and black cultivars when anthocyanin pigment accumulation starts in the skin cells and continues through the ripening phase. Hence, the characterization of skin tissue is an essential parameter for understanding grape ripening due to its key role in the development of compounds responsible for the quality of wine. Additionally, grape skins are of increasing interest because they are metabolically active during development and ripening, and may have an endocrinal function [18]. The accumulation of anthocyanin pigments in grape berry skin is an important determination of berry quality. Usually restricted to the skins of berries, these pigments provide essential cultivar differentiation for consumers and are implicated in the health benefits of grape berry. In grapes, pigment biosynthesis may be induced by light, particularly ultraviolet (UV) radiation, stress treatments, and enzymes [19–21]. Several studies have shown that anthocyanin biosynthetic enzymes are induced in coordination during the developmental process of grape berries [22–25]. In a study by Espley et al. [26], the authors suggested that, the expression of the genes encoding the biosynthetic enzymes is coordinately regulated by transcriptional regulatory proteins. Other studies have also reported that transcription factors (TFs) are involved in the regulation of genes in the anthocyanin biosynthetic pathway and the components of the regulatory complex controlling anthocyanin biosynthesis are conserved in higher plants [27]. According to Martin and Paz-Ares [28], MYB TFs have been shown to play an important role in transcriptional regulation of anthocyanins, indicating that plant MYBs control secondary metabolic pathways, plant development, and signal transduction [29]. The characteristics of these MYBs involve the structurally conserved DNA-binding domain consisting of single or multiple imperfect repeats and the ones associated with the anthocyanin pathway are of the two-repeat (R2R3) class [26].

The regulation of MYBs can also be specific to discrete subsets of structural genes acting early or late in the anthocyanin biosynthetic pathway [30]. In *Arabidopsis thaliana*, Stracke et al. [31] stated that, there are 126 R2R3 MYB TFs, which can be divided on the basis of their sequence into 24 subgroups. However, the R2R3 MYB factors that regulate anthocyanin biosynthesis have been shown to interact closely with basic helix-loop-helix (bHLH) TFs as reported by Mol et al. [32] and Winkel-Shirley [33]. MYB genes have also been studied in grapes [34–36] and determined to be involved in anthocyanin biosynthesis. Numerous studies [34,37–39], have determined that many of the white grape cultivars arose from multi-allelic mutations of the MYBA1 and MYBA2 genes, which control the final biosynthetic steps of anthocyanin synthesis. This is a glycosylation reaction mediated by the UDP-glucose flavonoid 3-*O*-glucosyltransferase (UFGT). Similarly, Deluc et al. [35], determined that MYB5a and MYB5B are involved in regulating several flavonoid biosynthesis steps. Six MYB transcription factors (MYBA1, MYBA2, MYB5a, MYB5b, MYBPA1, and MYBPA2) have been reported to be associated with the regulation of the structural genes in the flavonoid pathway [40]. Although studies of *MYB* genes as transcriptional regulators of the flavonoid pathway genes have been investigated in *V. vinifera* varieties, to date this has not been done in muscadine grapes. The control of anthocyanin accumulation in muscadine grapes is a key question in understanding and manipulating the color of the berries and in vitro cell lines. Identification of the factors that exert this control will provide tools for moderating the extent and distribution of anthocyanin-derived pigmentation in berries and cell tissues of muscadine grapes. In this study, we isolated, and cloned *MybA1*, and *MYBCS1* (*VrMYB5a*) from the berry skins of muscadine grapes for the first time. The gene sequences were then characterized using in silico analysis and eventually deposited in the GenBank. We also analyzed the transcriptional profile of these two genes in muscadine berry skin across three developmental stages using real-time PCR. We discovered a high coordination between transcription regulation of *VrMybA1*, and *VrMYBCS1* (*VrMYB5a*) with the accumulation profile of total anthocyanin in the development stages. The results from this study provide new knowledge on the characteristics of MYB genes in muscadine grapes.

2. Materials and Methods

2.1. Plant Materials, Growth Conditions, and Cell Culture Maintenance

Berry skins and red cell lines of muscadine grapes ("Noble" var.) were used in this study. Berries were harvested from the Florida A & M University vineyard at three different development stages (green, veraison, and physiological maturity). Veràison refers to the onset of ripening, while physiological maturity refers to berries with brix of 18 and above and ready for harvesting. Berries that were free of physical injuries and similar in size at every stage were chosen for mRNA extraction. They were washed with distilled water; the skins were peeled and immediately frozen with liquid nitrogen and stored at −80 °C until use. In vitro red cell cultures established from super-epidermal cells of red berries of muscadine grapes [41] were also used in this study. The cells were grown in a growth chamber at 23 °C under a white light (150 $\mu E \cdot m^{-2} \cdot s^{-1}$) with a 16 h light/8 h dark cycle. The developed callus produces anthocyanin that is red in color. Solid culture mediums were used to grow and maintain the cells. Grape cell cultures were maintained in B-5 media as previously described by reference [4,42]. The cells in solid media were sub-cultured every 30 days and cell suspensions were transferred to fresh liquid media every 12 days. To transfer the cells, light forceps were gently used to scoop off the top layers of the cells. They were gently spread onto new culture media in three medium size layers in each plate in order to give the cells enough room to multiply. In the liquid medium, approximately 2.5 mL of the cell suspension was transferred into a 25 mL Erlenmeyer flask with B-5 liquid medium and placed on a shaker at 135 rpm.

2.2. RNA Extraction, Gel Electrophoresis and cDNA Synthesis

Samples were prepared from the skins (green, veràison, and physiologically mature berries) as well as cell lines of "Noble" grape. Total RNA was isolated using the RNeasy Plant Mini Kit (Qiagen, Valencia, CA, USA) according to the manufacturer's protocol. RNA was quantified using Nanodrop 3300 (Thermo Scientific, Swedesboro, NJ, USA), and the inactivity was inspected by formaldehyde agarose gel electrophoresis. Purified RNA was treated with RNase-free DNAse 1, and immediately frozen to −20 °C. Formaldehyde gel electrophoresis (1% agarose) was used to evaluate the RNA quality. The gel apparatus (including the gel tray and comb) was treated with RNase Away™ (Molecular Bio-Products, Inc., San Diego, CA, USA) and rinsed with distilled water. Total RNA was used in primary gene expression profiling. The SuperScript First-strand Synthesis System for RT-PCR (Invitrogen, Carlsbad, CA, USA) was used to synthesize cDNA in a 20 mL reaction containing 1 mg of DNase I-treated total RNA, 20 mM Tris-HCl (pH 8.4), 50 mM KCl, 2.5 mM MgCl$_2$, 10 mM dithiothreitol, 0.5 mg oligo (dT), 0.5 mM each of dATP, dGTP, dCTP, and dTTP, and 200U SuperScript II Reverse Transcriptase. RNA, dNTPs, and oligo (dT) were mixed first, heated to 65 °C for 5 min, and placed on ice until the addition of the remaining reaction components. The reaction was incubated at 50 °C for 50 min, and terminated by heat inactivation at 85 °C for 5 min. The cDNA product was treated with 1 μL of Rnase H (Invitrogen) for 20 min at 37 °C. An identical reaction without the reverse transcriptase was performed to verify the absence of genomic DNA (no-RT control). The cDNA was stored at −20 °C until it was ready for use.

2.3. Isolation of the VrMybA1, and VrMYBCS1 (VrMYB5a) Genes

To isolate *MybA1*, and *MYBCS1* cDNA clones from muscadine grapes, primers were designed based on the basis of conserved amino acid sequences of several published *MybA1* and *MYBCS1* sequences from other plants to amplify the CDS region. Based on the sequences of published genes and specific sequences of *V. vinifera MybA1* and *MYBCS1* (NCBI accession numbers AB097923, and AY555190), two fragments were amplified from cDNA by the primers *MybA1-F*: 5′-CACCATGGAGAGCTTAGGAGTTAGA-3′ and *MybA1-R*: 5′-GATCAAGTGATTTACTTGTGT-3′, for *VrMybA1* and primers MYBCS1-F: 5′-CACCATGAGAAATCCGGCATCTGC-3′ and MYBCS1-R: 5′-GGGAGACATGGAGTGTTTTTGA-3′, for *VrMYBCS1*. cDNA synthesized from mRNA of the veràison berry skins was used as a template. A high fidelity polymerase (Promega, Madison, WI, USA) was used for PCR using the following program: 95 °C for 5 min, then 35 cycles of 95 °C for 50 s, 55 °C for 50 s, and 72 °C for 90 s; followed by elongation at 72 °C for 10 min. The PCR products were separated by 1% agarose gel electrophoresis, and strong bands were clearly detected at 530 bp and 760 bp, respectively representing *MybA1* and *MYBCS1*. The agarose gel slice containing the DNA fragment of interest was purified by DNA gel extraction kit (Qiagen, Valencia, CA, USA) according to the operator's manual.

2.4. Cloning and Sequencing of VrMybA1, and VrMYBCS1 (VrMYB5a) Genes

The PCR fragments of *VrMybA1*, and *VrMYBCS1* (*VrMYB5a*) genes were subsequently purified and cloned into pGEM-T Easy Vector (Promega). Vectors and PCR-amplified products were mixed and ligated overnight at 4 °C and transformed into *Escherichia coli* strain JM109. The putative recombinant plasmid-pGEM-*MybA1* and pGEM-*MYBCS1* were extracted for PCR analysis, and strong bands were clearly detected at 530 bp and 760 bp, respectively confirming the presence of the *MybA1* and *MYBCS1* inserts. The pGEM-*MybA1* and pGEM-MYBCS-1 plasmids were sequenced from both ends at Eurofins MWG/Operon (Huntsville, AL, USA). The sequences were compared with the genes in the GenBank database using BLAST program from the National Center for Biotechnology Information (NCBI), which indicated that the PCR product had 99% identity in the activity site with the reported *Vitis MybA1* and *MYBCS1*, which confirmed that the obtained genes are muscadine *MybA1* and *MYBCS1*.

2.5. Bioinformatics Analysis

The deduced amino acid sequence of *VrMybA1*, and *VrMYBCS-1* were aligned using the BioEdit Sequence Alignment Editor, version 5.0.9 (Department of Microbiology, North Carolina State University, Raleigh, NC, USA) [43]. Theoretical molecular weights (MW) and isoelectric points (pI) were calculated using the Compute pI/Mw tool (http://web.expasy.org/compute_pi/). Putative target localization of *VrMybA1*, and *VrMYBCS1* was predicted by using WoLF PSORT (http://psort.hgc.jp/form.html) [44]. The phylogenetic tree was constructed using an online program (http://www.phylogeny.fr/) [45]. The three-dimensional (3D) structure was built using the SWISS-MODEL program and illustrated with the PyMOL viewer.

2.6. Expression Pattern in Different Berry Stages and in Cell Cultures

The expression levels of *VrMybA1*, and *VrMYBCS1* genes at three stages of muscadine berry skin developments were determined by RT-PCR, using SYBR green method on a CFX96 real-time cycler (BIO-RAD, Hercules, CA, USA). Relative quantitative real-time PCR reactions were performed in a 96-well plate to monitor cDNA amplification, according to the manufacturer's protocol. As a control, a parallel amplification reaction of *Actin* (a housekeeping gene) was performed. Each primer set was designed based on the 3′-end cDNA sequence of the corresponding gene. The specific primers used for RT-PCR were as follows: for *VrMybA1* 5′-GGAATAGATCCCAGAACCCAC-3′ (Forward), and 5′-TGGTTGCTTCCAATTTCTCCC-3′ (Reverse), giving a product of 140 bp and *VrMYBCS-1* 5′-GAGGTTGATCTCATGATTAGG-3′ (Forward), and 5′-TGGAACTGAACCTCCTTTTTG-3′ (Reverse), giving a product of 180 bp; for *Actin* 5′-TAGAAGCACTTCCTGTGGAC-3′ (Forward) and 5′-GGAAATCACTGCACTTGCTC-3′ (Reverse), giving a product of 120 bp. Each PCR reaction (20 μL) contained 0.6 μL primer F, R (10 μM), 1 μL cDNA (10 ng), and 10 μL SsoAdvanced™ SYBR® Green Supermix (Bio-Rad, Hercules, CA, USA). The RT-PCR conditions were: 1 cycle at 95 °C for 3 min; 35 cycles at 95 °C for 10 s and 60 °C (*VrMybA1*, *VrMYBCS-1*, and Actin) for 30 s, followed by a melt cycle from 65.0 °C to 95.0 °C. Three replicates of all RT-PCR reactions were carried out on each sample. Amplification efficiency of all primers used was primarily determined prior to sample investigation. Relative expression values were firstly calculated as $2^{-\Delta CT}$, normalized against the internal control *Actin* gene. The maximal expression level of each gene observed served as a calibrator (1.0) respectively, and the rest was expressed as ratios in relation to the calibrator (relative expression ratio).

2.7. Analysis of Total Anthocyanin in the Berry Skins

Approximately 2 g of fresh tissue (callus and/or skins) were kept frozen at −80 °C, then homogenized with Bio Homogenizer (Biospec Products, Inc., Bartsville, OK, USA) at 5000 rpm for, 5 min, in 10 mL extraction solvent (Methanol: 1% HCl—1:1). It was then centrifuged for 15 min (4 °C) (Eppendorf 5804R, Swedesboro, NJ, USA) at 11,000 rpm. The supernatant was collected and the residue was homogenized and re-extracted two more times. The extract was combined and filtered through a 0.45 μm syringe filter and used for total anthocyanin assay.

The pH differential spectrophotometric method was used to measure total anthocyanin content [46]. Two portions of anthocyanin extracts were diluted (using pre-determined dilution factor) with potassium chloride buffer (0.025 M, pH 1.0) and sodium acetate buffer (0.4 M, pH 4.5), respectively. After 15 min, the absorbance of both dilutions was measured at 520 nm and 700 nm against water (Ultrospec 3100 Pro UV-Vis Spectrophotometer, GE Healthcare, Piscataway, NJ, USA). The corrected absorbance of the diluted sample and the monomeric anthocyanin pigment concentrations were calculated as described elsewhere [46]. The total monomeric anthocyanin concentration was expressed as mg cyanidin-3,5-diglucoside equivalents (molecular weight = 611.5254, molar absorptivity ε = 30175) per g dry weight (DW). Dry weights of muscadine berries skins and callus tissue were determined using MJ33 Compact Infrared Moisture Analyzer (Mettler Toledo, Greifensee, Switzerland).

3. Results

3.1. Isolation and Cloning of VrMybA1 and VrMYBCS1 Genes from Muscadine Berry Skin Tissues

Based on the alignment of homologous sequences from other plants, primer pairs were designed to amplify *VrMybA1* and *VrMYBCS1*. Using the homologous amplification method, these two regulatory genes were amplified successfully from the red grape skins. Agarose gel electrophoresis detection showed that the specific bands were 530 bp and 760 bp for *VrMybA1* and *VrMYBCS1*, respectively, which was consistent with the expected size based on the primer design. These bands were cloned and sequenced, and the exact sizes of *VrMybA1* and *VrMYBCS1* were determined to be 536 bp and 765 bp in length. After analysis and comparison of the sequences, we determined that the fragments had high similarity with known *MybA1* and *MYBCS1* genes from other plant species. The sizes obtained encoded protein sizes with 175 and 255 amino acid (AA) residues for VrMybA1 and VrMYBCS1, respectively (Figures 1 and 2). The trimmed nucleotide sequences of *VrMybA1* and *VrMYBCS1* genes coding for proteins were submitted to the NCBI/GenBank database under the following accession numbers: *VrMybA1* (KJ513437) and *VrMYBCS1* (KJ513438). The results of homology BLAST showed that both MYB genes cloned in this study shared high sequence identity, ranging from 97% to 99%, with cDNA sequences from other species in the *Vitis* family in the NCBI/GenBank database. The deduced amino acid sequences showed high similarity (Figure 3A,B), and the three-dimensional (3D) structure of muscadine *VrMybA1* and *VrMYBCS1* (Figure 4) shared 90.3% similarity with the template. This further facilitated the positive identification of *VrMybA1* and *VrMYBCS1*.

```
1     ATGATTAGGCTTCACAATTTGTTGGGGAACAGATGGTCCTTGATTGCGGGTAGGCTTCCA
1      M  I  R  L  H  N  L  L  G  N  R  W  S  L  I  A  G  R  L  P

61    GGGAGGACTGCTAATGATGTCAAGAACTATTGGCATAGTCACCACTTCAAAAAGGAGGTT
21     G  R  T  A  N  D  V  K  N  Y  W  H  S  H  H  F  K  K  E  V

121   CAGTTCCAGGAAGAAGGGAGAGATAAACCCCAAACACATTCTAAAACCAAAGCTATAAAG
41     Q  F  Q  E  E  G  R  D  K  P  Q  T  H  S  K  T  K  A  I  K

181   CCTCACCCTCACAAGTTCTCCAAAGCCTTGCCAAGGTTTGAACTAAAAACTACAGCTGTG
61     P  H  P  H  K  F  S  K  A  L  P  R  F  E  L  K  T  T  A  V

241   GATACTTTTGACACACAAGTCAGTACTTCCAGTAAGCCATCATCCACGTCACCACAACGG
81     D  T  F  D  T  Q  V  S  T  S  S  K  P  S  S  T  S  P  Q  R

301   AATGATGACATCATATGGTGGGAAAGCCTGTTAGCTGAGCATGCTCCAATGGATCAAGAA
101    N  D  D  I  I  W  W  E  S  L  L  A  E  H  A  P  M  D  Q  E

361   ACTGACTTTTCGGCTTCTGGAGAGATGCTTATCGCAAGCCTCAGGACAGAAGAAACTGCA
121    T  D  F  S  A  S  G  E  M  L  I  A  S  L  R  T  E  E  T  A

421   ACACAGAAAAAGGGACCCATGGATGGTATGATTGAACAAATCCAGGGAGGTGAGGGTGAT
141    T  Q  K  K  G  P  M  D  G  M  I  E  Q  I  Q  G  G  E  G  D

481   TTTCCATTTGATGTGGGCTTCTGGGATACACCCAACACACAAGTAAATCACTTGAT
161    F  P  F  D  V  G  F  W  D  T  P  N  T  Q  V  N  H  L
```

Figure 1. The complete cDNA sequence and amino acid sequence of the protein encoded by *MybA1* (GenBank accession number: KJ513437).

```
1     ATGAATTATCTTCGGCCGTCAGTGAAGCGCGGCCAGATAGCTCCCGATGAGGAAGATCTC
1      M  N  Y  L  R  P  S  V  K  R  G  Q  I  A  P  D  E  E  D  L

61    ATTCTTCGCCTCCATCGCCTGCTCGGTAACAGGTGGTCTCTGATTGCCGGAAGGATCCCG
21     I  L  R  L  H  R  L  L  G  N  R  W  S  L  I  A  G  R  I  P

121   GGGCGTACAGACAATGAGATCAAGGACTACTGGAACACCCATCTCAGCAAGAAACTCATC
41     G  R  T  D  N  E  I  K  D  Y  W  N  T  H  L  S  K  K  L  I

181   AGCCAAGGAATAGATCCCAGAACCCACAAGCCACTAAACCCTAAACCTAATCCATCACCA
61     S  Q  G  I  D  P  R  T  H  K  P  L  N  P  K  P  N  P  S  P

241   GATGTTAATGCTCCTGTGTCAAAATCAATTCCAAATGCAAACCCTAACCCTAGTTCTTCC
81     D  V  N  A  P  V  S  K  S  I  P  N  A  N  P  N  P  S  S  S

301   CGGGTGGGAGAAATTGGAAGCAACCATGAGGTCAAGGAGATTGAAAGTAATGAAAATCAC
101    R  V  G  E  I  G  S  N  H  E  V  K  E  I  E  S  N  E  N  H

361   AAGGAGCCGCCTAACCTGGATCAGTATCACAGTCCACTTGCGGCCGATAGCAATGAGAAT
121    K  E  P  P  N  L  D  Q  Y  H  S  P  L  A  A  D  S  N  E  N

421   TGGCAAAGCGCAGATGGGTTGGTAACGGGACTACAAAGCACCCATGGTACCAGCAACGAT
141    W  Q  S  A  D  G  L  V  T  G  L  Q  S  T  H  G  T  S  N  D

481   GACGAAGACGATATCGGGTTCTGCAACGACGATACATTCTTCTTCATTTTTGAATTCTTTG
161    D  E  D  D  I  G  F  C  N  D  D  T  F  S  S  F  L  N  S  L

541   ATTAACGAGGATGTGTTTGGAAATCATAATCATCATCAGCAGCAGCAACAGCAACAGCAG
181    I  N  E  D  V  F  G  N  H  N  H  H  Q  Q  Q  Q  Q  Q  Q  Q

601   CTGCAGCAGCTGCAGCAGCCATCTAATGTGATTGCACCATTGCCCCACCCAGCAATTTCT
201    L  Q  Q  L  Q  Q  P  S  N  V  I  A  P  L  P  H  P  A  I  S

661   GTGCAGGCCACCTTCAGTAGTAGCCCTAGAACTGTCTGGGAACCTGCTGCACTAACATCT
221    V  Q  A  T  F  S  S  S  P  R  T  V  W  E  P  A  A  L  T  S

721   ACATCGGCTCCTTTAGTCCACGATCAAAAACACTCCATGTCTCCC
241    T  S  A  P  L  V  H  D  Q  K  H  S  M  S  P
```

Figure 2. The complete cDNA sequence and amino acid sequence of the protein encoded by *MYBCS1* (GenBank accession number: KJ513438).

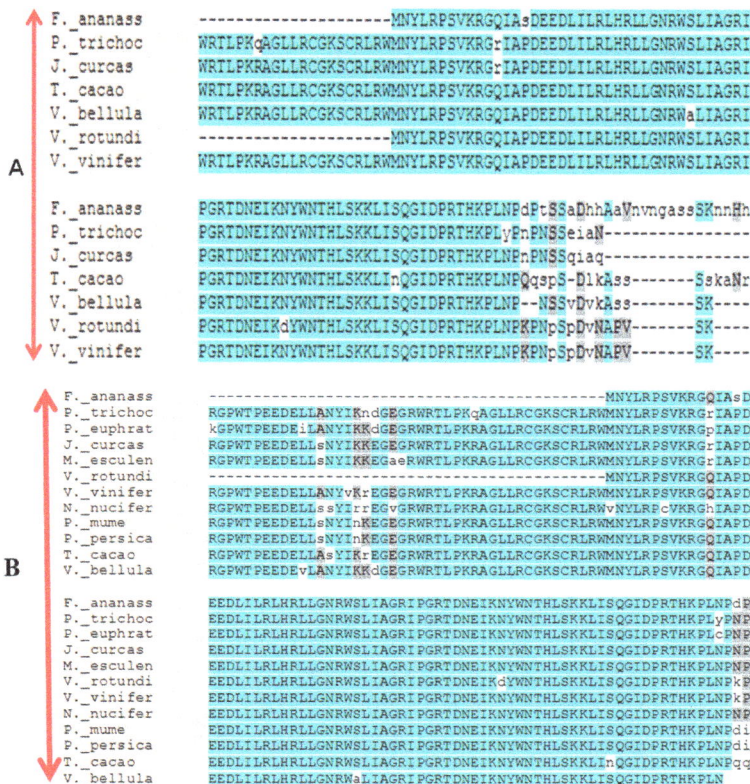

Figure 3. Amino acid sequence alignment of MybA1 (**A**); and MYBCS1 (**B**) proteins from different plant species. Residues highlighted in blue represent identical and similar amino acids, respectively. The alignment was performed using multi-align software as defined in the Materials and Methods section.

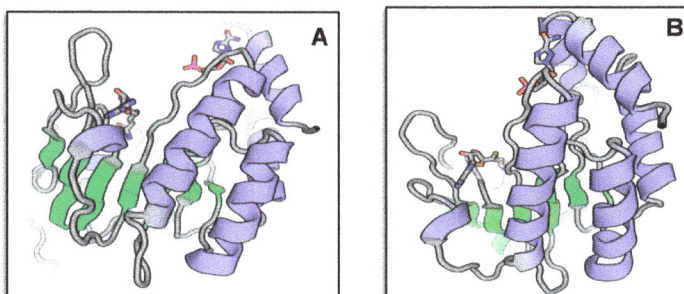

Figure 4. The computational modeled three dimensional structure of muscadine *MybA1* (**A**); and *MYBCS-1* (**B**).

3.2. Sequence Analysis of VrMybA1 and VrMYBCS1

Sequence analysis indicated that *VrMybA1* and *VrMYBCS1* were 536 bp and 765 bp in length with 121 and 253 bp on the 5′- and 3′-untranslated regions (UTRs), respectively for *VrMybA1* and 186 bp and

186 bp on the 5′- and 3′-UTRs, respectively for *VrMYBCS1*. The cDNA contained an open reading frame (ORF) of 536 bp for *VrMybA1* and 765 bp for *VrMYBCS-1* encoding 178 AA and 255 AA, respectively (Figures 1 and 2). The deduced molecular mass (MW) of VrMybA1 protein was 20.2 kDA, with an isoelectric point (pI) of 6.3. For VrMYBCS-1 protein, the MW was 28.3 kDa with the pI of 5.65. SMART program (http://smart.embl-heidelberg.de/), was used to predict functional sites and results indicated there was a SANT domain, implicated in chromatin-remodeling and transcription regulation, between Trp85 and Phe135 for VrMybA1 (*E*-value 7.24×10^{-2}) and for VrMYBCS-1 (*E*-value 2.29×10^{-13}). The domain SANT/MYB of *VrMybA1* and *VrMYBCS-1* and their orthologues were compared with that of typical MYB domains found in plant MYB domain proteins, and with SANT domain often presented in proteins participating in response to anthocyanin accumulation. There was a broader search for genes with high homology to *VrMybA1* and *VrMYBCS-1* protein in other plant species and we found that *VrMybA1* and *VrMYBCS-1* contained a highly conserved DNA binding domain that was very similar to the DNA binding domains of other plant MYBs. The R2R3 imperfect repeats that bind to the target DNA sequences and are highly conserved among R2R3-MYB proteins were contained at the amino terminal. In *VrMYBCS-1*, there was a shorter sequence that is related to maize transcription factor activator C1 [47], and is needed to interact with a basic helix-loop-helix cofactor [35]. In addition to the C1 motif, a Gln-rich domain, QQQQQQQQLQQLQQP was identified. This motif resembles that found in *VvMYB5a* by reference [35]. Therefore, this study confirms that we have positively isolated and identified *VrMybA1* and *VrMYBCS-1* in muscadine grapes. This is very important in understanding the metabolic flux in the anthocyanin biosynthetic pathway of muscadine grapes.

A phylogenetic tree based on amino acid alignment of *VrMybA1* and *VrMYBCS-1* and related proteins with SANT (Swi3, Ada2, N-Cor, and TFIIIB) and MYB domains was constructed using the neighbor-joining method of MEGA 4.0 [48]. The dendrograms indicated that VrMybA1 and VrMYBCS1 belong to a distinct cluster of MYB proteins from *Vitis* species (Figure 5A,B). MybA1 from different organisms were divided into three groups: I, II, and III (Figure 5A), and MYBCS1 was also divided into three groups: I, II, and III (Figure 5B). VrMybA1 from *V. rotundifolia* belonged to the subgroup II that is close to the *V. vinifera* and MYBCS1 from *V. rotundifolia* belonged to the subgroup II, which is also in the same clade as *V. vinifera*. Even though members of each subgroup have close relationships, the study of this relationship through phylogenetic analysis has enabled us to visualize how evolution has produced changes in the MYB gene family.

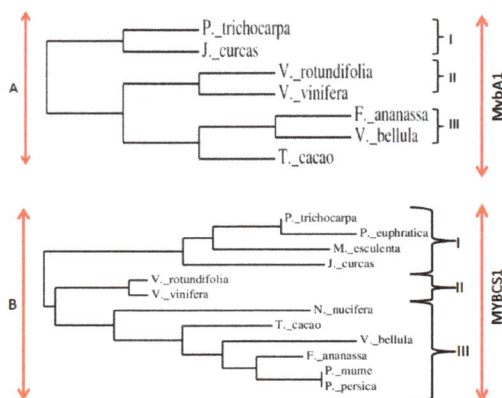

Figure 5. Comparison of the deduced amino acid sequences of R2R3-type MYB transcription factors from higher plants. (**A**) A phylogenetic tree for plant *MybA1* transcription factors; (**B**) A phylogenetic tree for plant *MYBCS1* transcription factors including the one isolated from muscadine grapes (*V. rotundifolia*) in the present study.

3.3. Expression of VrMybA1 and VrMYBCS1 Genes in Berry Skins Are Influenced by Physiological Changes

Real-time PCR was carried out to investigate the expression patterns of *VrMybA1* and *VrMYBCS1* genes induced by the physiological changes during grape berry developmental (Figure 6A–D). Real-time PCR analysis showed that the relative expression of *MybA1* (Figure 6I) and *MYBCS1* (Figure 6II) was significant among the tested samples. Expressions of these genes were not detected at the green stage. But at veràison, the skin color starts to change from green-to-red and the expressions of *VrMybA1* and *VrMYBCS-1* genes were already detectable in the red skin tissues. In the skins of physiologically mature berries, red pigments were actively developed, and levels of *VrMybA1* and *VrMYBCS-1* gene transcripts were greatly induced. To determine the contribution of *VrMybA1* and *VrMYBCS-1* in the flavonoid pathway of *V. rotundifolia*, the expression levels of these genes were analyzed in the in vitro cell cultures. The data obtained indicated that the expression levels of *MybA1* and *MYBCS-1* were significantly higher in the in vitro red cell lines as compared to that observed in the green, veràison, and mature skin tissues (Figure 6). Thus, this study confirms that *MybA1* and *MYBCS-1* play an important role in the accumulation of anthocyanin in *V. rotundifolia*.

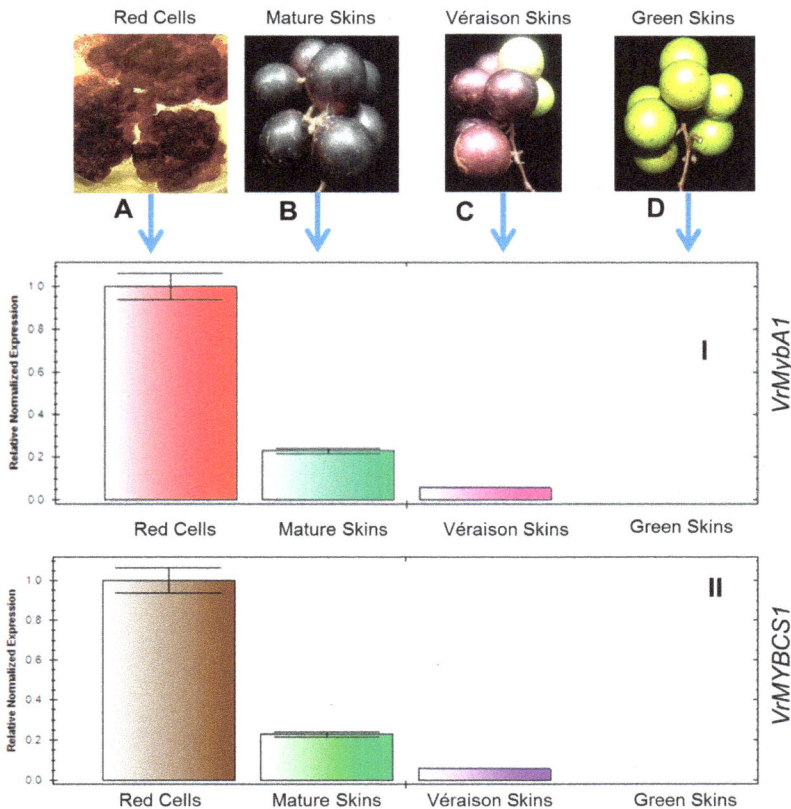

Figure 6. Plant samples used in the experiment (**A**) (Red cells); (**B**) (Mature berries); (**C**) (Veràison berries); and (**D**) (Green berries) and mRNA expression levels of *VrMybA1* (**I**); and *VrMYBCS-1* (**II**) genes in different developmental stages of muscadine grape berry skins. The values are expressed relative to a level of transcription (**I**) the veràison, which was set as 0.1. Values are the means of three replicates ± SE.

3.4. Anthocyanin Accumulation in the Berry Skin and Cell Cultures

Total anthocyanin was measured in the three phenotypes of berry skins (green, veràison, and physiologically mature). Total anthocyanin content was high in the in vitro red cell cultures and at a late stage of berry development or physiologically mature berry skins (Figure 7). The analysis indicated that anthocyanin accumulates progressively during berry development. It starts to accumulate on the berry skin at veràison and increases until it maximizes at physiological maturity. Comparatively, anthocyanin was low or non-existent at the green developmental stage (Figure 7). These results are consistent with the results shown in Figure 6I,II, where the red and green skin color differences were observed progressively in different stages of berry development. From these results, we can deduce that the accumulation of anthocyanin pigments in berry skin is likely influenced by the progressive expression of *VrMybA1* and *VrMYBCS-1*.

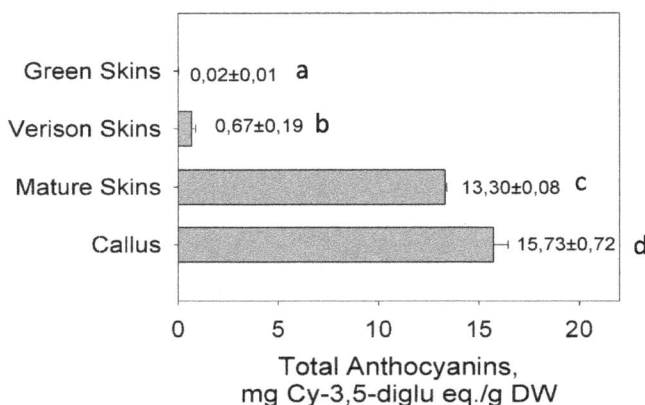

Figure 7. Analysis of anthocyanin accumulation in in vitro cell cultures (Callus), as well as in different development stages of muscadine berry skins (Mature, Veriason, and Green skins). Mean values with different small letters (**a**–**d**) are significantly different at ($p = 0.01$). The accumulation pattern of anthocyanins follows the pattern of expression of *VrMybA1*, and *VrMYBCS-1*.

4. Discussion

MYB represent the largest number of proteins with active functions in numerous plant species. Thus far, 190 have been deduced in *Arabidopsis*, 156 in rice, and more than 200 have been identified in maize [49–51]. MYB transcriptional regulators contain conserved DNA-binding domains that are usually composed of one, two, or three imperfect 51 or 52 residue repeats (R1, R2, and R3). Each repeat encodes three α-helices, with the second and third helices forming a helix–turn–helix (HTH) structure when bound to DNA [52,53]. In plants, the overwhelming majority of MYB proteins are classified into a subfamily characterized by the presence of R2R3 motifs.

The genetics and biochemistry of anthocyanins and flavonoid biosynthesis in plant organ pigmentation are well established in model species. Although the molecular and biochemical characterization of *MybA1* and *MYBCS-1* or analogous genes in plants have been studied, most of the studies were carried out in model plant species, such as *Arabidopsis*, tobacco, and rice. However, *MybA1* and *MYBCS-1* or its analogues in muscadine grapes (*Vitis rotundifolia*), have not been reported before. The focus of this study was to investigate the existence and expression of *MybA1* and *MYBCS-1* in muscadine grapes and their possible involvement in anthocyanin biosynthesis. The cloning, characterization and expression of *MybA1* and *MYBCS-1* genes from muscadine grapes will create the possibility of elucidating the biosynthetic pathway in *V. rotundifolia*, which is now under intensive investigation in our laboratory.

Although regulatory genes as well as structural genes in flavonoid pathway have been isolated in other plant species, there are currently no reports of *VrMybA1* and *VrMYBCS-1* in *V. rotundifolia*. The results indicated that *VrMybA1* and *VrMYBCS-1* are key enzymes in the anthocyanin-pigmentation pathway of *V. rotundifolia*, and the two genes are responsible for the formation of anthocyanin in the berry skins as well as the in vitro cell lines of *V. rotundifolia*. Flavonoids are responsible for many medical applications of grapes; therefore, the manipulation of the flavonoid biosynthetic pathway in muscadine grape cell cultures would provide an alternative to harvesting bioactive compounds from the grape cells [3,4]. *MybA1* and *MYBCS-1*, are key enzymes of the anthocyanin biosynthetic pathway, and can regulate the structural genes and catalyze the reaction(s) from the colorless leucoanthocyanidins to the colored anthocyanidins. The present research on isolation and characterization analysis of *MybA1* and *MYBCS-1* from muscadine grapes will provide new research opportunities for the overall metabolic flux control toward the targeted products using genetic engineering strategies.

5. Conclusions

In the present study, we successfully isolated and characterized two MYB genes (*VrMybA1* and *VrMYCS-1*) from muscadine grapes and analyzed their expression profiles in different stages of berry development as well as in in vitro cell lines for the first time. Phylogenetic relationship constructed based on the putative amino acid sequences demonstrated that *VrMybA1* and *VrMYCS-1* were most closely related to *V. vinifera*, among the surveyed plant species. Multiple alignments of amino acid sequences of *VrMybA1* and *VrMYCS-1* in relation to others showed that they have many active sites that are well-conserved in different plant species. Our data revealed that transcripts levels of *VrMybA1* and *VrMYCS-1* were influenced by the accumulation of anthocyanins in the berry skins and cell lines, implying that they play a vital role in anthocyanin accumulation. Therefore, *VrMybA1* and *VrMYCS-1* could be considered as potential targets in genetic engineering for producing transgenic plants with improved anthocyanin accumulation. To gain greater insight into the functions of *VrMybA1* and *VrMYCS-1* in anthocyanin accumulation in muscadine grapes, further research will focus on the analysis of *VrMybA1* and *VrMYCS-1* in transgenic *V. rotundifolia* plants and cell lines.

Acknowledgments: The research was carried out with the financial support of USDA/NIFA/AFRI Plant Biochemistry Program Grant #2009-03127 and USDA/NIFA/1890 Capacity Building Grant #2010-02388.

Author Contributions: Anthony Ananga and Violeta Tsolova conceived and designed the experiments; Lillian Oglesby and Anthony Ananga conducted the experiments, James Obuya and Ernst Cebert helped in data analysis, Lillian Oglesby, Anthony Ananga, and Violeta Tsolova wrote the paper with inputs from Joel Ochieng and Ernst Cebert.

Conflicts of Interest: The authors declare no conflict of interest.

References

1. Georgiev, V.; Ananga, A.; Tsolova, V. Recent advances and uses of grape flavonoids as nutraceuticals. *Nutrients* **2014**, *6*, 391–415. [CrossRef] [PubMed]
2. Georgiev, V.; Ananga, A.; Tsolova, V. Dietary supplements/nutraceuticals made from grapes and wines. In *Wine Safety, Consumer Preference, and Human Health*; Springer: Cham, Switzerland, 2016; pp. 201–227.
3. Ananga, A.; Georgiev, V.; Tsolova, V. Manipulation and engineering of metabolic and biosynthetic pathway of plant polyphenols. *Curr. Pharm. Des.* **2013**, *19*, 6186–6206. [CrossRef] [PubMed]
4. Ananga, A.; Phills, B.; Ochieng, J.; Georgiev, V.; Tsolova, V. *Production of Anthocyanins in Grape Cell Cultures: A Potential Source of Raw Material for Pharmaceutical, Food, and Cosmetic Industries*; InTech Open Access Publisher: Rijeka, Croatia, 2013.
5. Samuelian, S.K.; Camps, C.; Kappel, C.; Simova, E.P.; Delrot, S.; Colova, V.M. Differential screening of overexpressed genes involved in flavonoid biosynthesis in North American native grapes: "Noble" *muscadinia* var. and "Cynthiana" *aestivalis* var. *Plant Sci.* **2009**, *177*, 211–221. [CrossRef]

6.   Sandhu, A.K.; Gu, L. Antioxidant capacity, phenolic content, and profiling of phenolic compounds in the seeds, skin, and pulp of *Vitis rotundifolia* (muscadine grapes) as determined by HPLC-DAD-ESI-MS^n. *J. Agric. Food Chem.* **2010**, *58*, 4681–4692. [CrossRef] [PubMed]

7.   Huang, Z.; Wang, B.; Williams, P.; Pace, R.D. Identification of anthocyanins in muscadine grapes with HPLC-ESI-MS. *LWT-Food Sci. Technol.* **2009**, *42*, 819–824. [CrossRef]

8.   Hopkins, D.; Purcell, A. *Xylella fastidiosa*: Cause of Pierce's disease of grapevine and other emergent diseases. *Plant Dis.* **2002**, *86*, 1056–1066. [CrossRef]

9.   Fry, S.; Milholland, R. Response of resistant, tolerant, and susceptible grapevine tissues to invasion by the Pierce's disease bacterium, *Xylella fastidiosa*. *Phytopathology* **1990**, *80*, 66–69. [CrossRef]

10.   Brownleader, M.D.; Jackson, P.; Mobasheri, A.; Pantelides, A.T.; Sumar, S.; Trevan, M.; Dey, P.M. Molecular aspects of cell wall modifications during fruit ripening. *Crit. Rev. Food Sci. Nutr.* **1999**, *39*, 149–164. [CrossRef] [PubMed]

11.   Robinson, S.P.; Davies, C. Molecular biology of grape berry ripening. *Aust. J. Grape Wine Res.* **2000**, *6*, 175–188. [CrossRef]

12.   Brummell, D.A.; Dal Cin, V.; Crisosto, C.H.; Labavitch, J.M. Cell wall metabolism during maturation, ripening and senescence of peach fruit. *J. Exp. Bot.* **2004**, *55*, 2029–2039. [CrossRef] [PubMed]

13.   Deytieux, C.; Geny, L.; Lapaillerie, D.; Claverol, S.; Bonneu, M.; Donèche, B. Proteome analysis of grape skins during ripening. *J. Exp. Bot.* **2007**, *58*, 1851–1862. [CrossRef] [PubMed]

14.   Deytieux-Belleau, C.; Geny, L.; Roudet, J.; Mayet, V.; Donèche, B.; Fermaud, M. Grape berry skin features related to ontogenic resistance to *Botrytis cinerea*. *Eur. J. Plant Pathol.* **2009**, *125*, 551–563. [CrossRef]

15.   Antolín, M.C.; Baigorri, H.; Luis, I.D.; Aguirrezábal, F.; Geny, L.; Broquedis, M. ABA during reproductive development in non-irrigated grapevines (*Vitis vinifera* L. cv. *Tempranillo*). *Aust. J. Grape Wine Res.* **2003**, *9*, 169–176. [CrossRef]

16.   Geny, L.; Deytieux, C.; Donèche, B. Importance of Hormonal Profile on the Onset of Ripening in Grape Berries of *Vitis vinifera* L. In Proceedings of the V International Postharvest Symposium 682, Verona, Italy, 6 June 2004; pp. 99–106.

17.   Okamoto, G.; Kuwamura, T.; Hirano, K. Effects of water deficit stress on leaf and berry ABA and berry ripening in Chardonnay grapevines (*Vitis vinifera*). *VITIS J. Grapevine Res.* **2015**, *43*, 15–17.

18.   Kanellis, A.; Roubelakis-Angelakis, K. Grape. In *Biochemistry of Fruit Ripening*; Springer: Dordrecht, The Netherlands, 1993; pp. 189–234.

19.   Keller, M.; Hrazdina, G. Interaction of nitrogen availability during bloom and light intensity during veràison. II. Effects on anthocyanin and phenolic development during grape ripening. *Am. J. Enol. Vitic.* **1998**, *49*, 341–349.

20.   Berli, F.J.; Moreno, D.; Piccoli, P.; Hespanhol-Viana, L.E.; Silva, M.F.; Bressan-Smith, R.; Cavagnaro, J.B.; Bottini, R. Abscisic acid is involved in the response of grape (*Vitis vinifera* L.) cv. Malbec leaf tissues to ultraviolet-B radiation by enhancing ultraviolet-absorbing compounds, antioxidant enzymes and membrane sterols. *Plant Cell Environ.* **2010**, *33*, 1–10. [PubMed]

21.   Takos, A.M.; Jaffé, F.W.; Jacob, S.R.; Bogs, J.; Robinson, S.P.; Walker, A.R. Light-induced expression of a *MYB* gene regulates anthocyanin biosynthesis in red apples. *Plant Physiol.* **2006**, *142*, 1216–1232. [CrossRef] [PubMed]

22.   Castellarin, S.D.; Pfeiffer, A.; Sivilotti, P.; Degan, M.; Peterlunger, E.; Di Gaspero, G. Transcriptional regulation of anthocyanin biosynthesis in ripening fruits of grapevine under seasonal water deficit. *Plant Cell Environ.* **2007**, *30*, 1381–1399. [CrossRef] [PubMed]

23.   Bogs, J.; Jaffé, F.W.; Takos, A.M.; Walker, A.R.; Robinson, S.P. The grapevine transcription factor VvMYBPA1 regulates proanthocyanidin synthesis during fruit development. *Plant Physiol.* **2007**, *143*, 1347–1361. [CrossRef] [PubMed]

24.   Hall, J.; Ananga, A.; Georgiev, V.; Ochieng, J.; Cebert, E.; Tsolova, V. Molecular Cloning, Characterization, and Expression Analysis of Flavanone 3-Hydroxylase (F3H) Gene during Muscadine Grape Berry Development. *J. Biotechnol. Biomater.* **2015**, *5*, 1–7.

25.   Corbiere, P.; Ananga, A.; Ochieng, J.; Cebert, E.; Tsolova, V. Gene Expression and Molecular Architecture Reveals UDP-Glucose: Flavonoid-3-*O*-Glucosyltransferase *UFGT* as a Controller of Anthocyanin Production in Grapes. *Jacobs J. Biotechnol. Bioeng.* **2015**, *2*. [CrossRef]

26. Espley, R.V.; Hellens, R.P.; Putterill, J.; Stevenson, D.E.; Kutty-Amma, S.; Allan, A.C. Red colouration in apple fruit is due to the activity of the MYB transcription factor, MdMYB10. *Plant J.* **2007**, *49*, 414–427. [CrossRef] [PubMed]

27. Holton, T.A.; Cornish, E.C. Genetics and biochemistry of anthocyanin biosynthesis. *Plant Cell* **1995**, *7*, 1071–1083. [CrossRef] [PubMed]

28. Martin, C.; Paz-Ares, J. MYB transcription factors in plants. *Trends Genet.* **1997**, *13*, 67–73. [CrossRef]

29. Jin, H.; Martin, C. Multifunctionality and diversity within the plant MYB-gene family. *Plant Mol. Biol.* **1999**, *41*, 577–585. [CrossRef] [PubMed]

30. Davies, K.M.; Schwinn, K.E. Transcriptional regulation of secondary metabolism. *Funct. Plant Biol.* **2003**, *30*, 913–925. [CrossRef]

31. Stracke, R.; Werber, M.; Weisshaar, B. The R2R3-MYB gene family in *Arabidopsis thaliana*. *Curr. Opin. Plant Biol.* **2001**, *4*, 447–456. [CrossRef]

32. Mol, J.; Jenkins, G.; Schäfer, E.; Weiss, D.; Walbot, V. Signal perception, transduction, and gene expression involved in anthocyanin biosynthesis. *Crit. Rev. Plant Sci.* **1996**, *15*, 525–557. [CrossRef]

33. Winkel-Shirley, B. Flavonoid biosynthesis: A colorful model for genetics, biochemistry, cell biology, and biotechnology. *Plant Physiol.* **2001**, *126*, 485–493. [CrossRef] [PubMed]

34. Kobayashi, S.; Goto-Yamamoto, N.; Hirochika, H. Retrotransposon-induced mutations in grape skin color. *Science* **2004**, *304*, 982. [CrossRef] [PubMed]

35. Deluc, L.; Barrieu, F.; Marchive, C.; Lauvergeat, V.; Decendit, A.; Richard, T.; Carde, J.P.; Mérillon, J.M.; Hamdi, S. Characterization of a grapevine R2R3-MYB transcription factor that regulates the phenylpropanoid pathway. *Plant Physiol.* **2006**, *140*, 499–511. [CrossRef] [PubMed]

36. Deluc, L.; Bogs, J.; Walker, A.R.; Ferrier, T.; Decendit, A.; Merillon, J.M.; Robinson, S.P.; Barrieu, F. The transcription factor VvMYB5b contributes to the regulation of anthocyanin and proanthocyanidin biosynthesis in developing grape berries. *Plant Physiol.* **2008**, *147*, 2041–2053. [CrossRef] [PubMed]

37. Lijavetzky, D.; Ruiz-García, L.; Cabezas, J.A.; De Andrés, M.T.; Bravo, G.; Ibáñez, A.; Carreño, J.; Cabello, F.; Ibáñez, J.; Martínez-Zapater, J.M. Molecular genetics of berry colour variation in table grape. *Mol. Genet. Genom.* **2006**, *276*, 427–435. [CrossRef] [PubMed]

38. Walker, A.R.; Lee, E.; Robinson, S.P. Two new grape cultivars, bud sports of Cabernet Sauvignon bearing pale-coloured berries, are the result of deletion of two regulatory genes of the berry colour locus. *Plant Mol. Biol.* **2006**, *62*, 623–635. [CrossRef] [PubMed]

39. Walker, A.R.; Lee, E.; Bogs, J.; McDavid, D.A.; Thomas, M.R.; Robinson, S.P. White grapes arose through the mutation of two similar and adjacent regulatory genes. *Plant J.* **2007**, *49*, 772–785. [CrossRef] [PubMed]

40. Ali, M.B.; Howard, S.; Chen, S.; Wang, Y.; Yu, O.; Kovacs, L.G.; Qiu, W. Berry skin development in Norton grape: Distinct patterns of transcriptional regulation and flavonoid biosynthesis. *BMC Plant Biol.* **2011**, *11*. [CrossRef] [PubMed]

41. Colova, V. Synchronized Strains of Subepidermal Cells of Muscadine (muscadine sp.) Grapevine Pericarp for Use as a Sourse of Flavonoids (nutraceuticals). U.S. Patent No. US 2011/0054195 A1, 3 March 2011.

42. Gollop, R.; Even, S.; Colova-Tsolova, V.; Perl, A. Expression of the grape dihydroflavonol reductase gene and analysis of its promoter region. *J. Exp. Bot.* **2002**, *53*, 1397–1409. [CrossRef] [PubMed]

43. Hall, T.A. BioEdit: A user-friendly biological sequence alignment editor and analysis program for Windows 95/98/NT. In *Nucleic Acids Symposium Series*; Oxford University Press: Oxford, UK, 1999; pp. 95–98.

44. Horton, P.; Park, K.J.; Obayashi, T.; Fujita, N.; Harada, H.; Adams-Collier, C.J.; Nakai, K. WoLF PSORT: Protein localization predictor. *Nucleic Acids Res.* **2007**, *35*, W585–W587. [CrossRef] [PubMed]

45. Dereeper, A.; Guignon, V.; Blanc, G.; Audic, S.; Buffet, S.; Chevenet, F.; Dufayard, J.F.; Guindon, S.; Lefort, V.; Lescot, M.; et al. Phylogeny. fr: Robust phylogenetic analysis for the non-specialist. *Nucleic Acids Res.* **2008**, *36*, W465–W469. [CrossRef] [PubMed]

46. Giusti, M.M.; Wrolstad, R.E. Characterization and measurement of anthocyanins by UV-visible spectroscopy. *Curr. Protoc. Food Anal. Chem.* **2001**. [CrossRef]

47. Grotewold, E.; Sainz, M.B.; Tagliani, L.; Hernandez, J.M.; Bowen, B.; Chandler, V.L. Identification of the residues in the Myb domain of maize C1 that specify the interaction with the bHLH cofactor R. *Proc. Natl. Acad. Sci. USA* **2000**, *97*, 13579–13584. [CrossRef] [PubMed]

48. Tamura, K.; Peterson, D.; Peterson, N.; Stecher, G.; Nei, M.; Kumar, S. MEGA5: Molecular evolutionary genetics analysis using maximum likelihood, evolutionary distance, and maximum parsimony methods. *Mol. Biol. Evol.* **2011**, *28*, 2731–2739. [CrossRef] [PubMed]

49. Riechmann, J.L.; Heard, J.; Martin, G.; Reuber, L.; Jiang, C.Z.; Keddie, J.; Adam, L.; Pineda, O.; Ratcliffe, O.J.; Samaha, R.R.; et al. Arabidopsis transcription factors: Genome-wide comparative analysis among eukaryotes. *Science* **2000**, *290*, 2105–2110. [CrossRef] [PubMed]

50. Goff, S.A.; Ricke, D.; Lan, T.H.; Presting, G.; Wang, R.; Dunn, M.; Glazebrook, J.; Sessions, A.; Oeller, P.; Varma, H.; et al. A draft sequence of the rice genome (*Oryza sativa* L. ssp. *japonica*). *Science* **2002**, *296*, 92–100. [CrossRef] [PubMed]

51. Dias, A.P.; Braun, E.L.; McMullen, M.D.; Grotewold, E. Recently duplicated maize *R2R3 MyB* genes provide evidence for distinct mechanisms of evolutionary divergence after duplication. *Plant Physiol.* **2003**, *131*, 610–620. [CrossRef] [PubMed]

52. Gabrielsen, O.S.; Sentenac, A.; Fromageot, P. Specific DNA binding by c-Myb: Evidence for a double helix-turn-helix-related motif. *Science* **1991**, *253*, 1140–1143. [CrossRef] [PubMed]

53. Ogata, K.; Morikawa, S.; Nakamura, H.; Sekikawa, A.; Inoue, T.; Kanai, H.; Sarai, A.; Ishii, S.; Nishimura, Y. Solution structure of a specific DNA complex of the Myb DNA-binding domain with cooperative recognition helices. *Cell* **1994**, *79*, 639–648. [CrossRef]

MDPI AG

St. Alban-Anlage 66

4052 Basel, Switzerland

Tel. +41 61 683 77 34

Fax +41 61 302 89 18

http://www.mdpi.com

Antioxidants Editorial Office

E-mail: antioxidants@mdpi.com

http://www.mdpi.com/journal/antioxidants